Aid vs. Trade

U.S. food aid has saved millions of lives made destitute by the failure of pastures and fields. Surplus, subsidized grains bought by the U.S. government from American farmers and shipped halfway around the world can also undercut productive farmers in developing countries, where warehouses may be overflowing with wheat, corn, sorghum, beans, or cotton.

Culture of Monoculture

Over the past decade, an estimated 1.4 million Mexicans have been forced off their lands in search for work in their own country or north of the border. Free trade policies and the "dumping" of cheap U.S. subsidized corn have devastated Mexico's traditional agriculture. Despite $3 to $5 billion paid each year to corn growers in the form of direct payments, marketing loans, insurance, disaster relief, and other programs, U.S. family farmers are also struggling.

Water Wars and Animal Gulags

In September 2002, the Bureau of Reclamation diverted water from the Klamath River to maintain a heavily subsidized farming industry. At least 33,000 adult fish, primarily Chinook salmon, and an equal number of juveniles, fell victim to the largest salmon die-off in history. On a modern confinement animal feedlot operation, 16,000 dairy cows make twice-daily journeys from feeding stall to automated milking pen.

French Bakery

Modern Food Chain

Most subsidized crops enter our diets indirectly—and almost invisibly—as feed or processed ingredients. The U.S. Surgeon General reports that Americans spend $110 billion annually on illnesses caused by obesity, with taxpayers footing an increasingly large percentage of these costs.

Changes in farm and food policy can help to bring these preventable diseases under control.

At the same time, nearly 40 million Americans, 12 percent of all houseolds, confront food insecurity, meaning that they often experience hunger or need to skip meals to get by. Many are children.

It's What's for Lunch

The food industry spends $15 billion per year marketing to children. The Federal School Lunch Program spends only $7 billion to feed our children in the public schools. The Federal School Lunch Reimbursement is $2.32 per meal. Approximately 95 cents is spent on food and 97 cents on labor.

FOODFIGHT

THE CITIZEN'S GUIDE TO A FOOD AND FARM BILL

Daniel Imhoff

Foreword by Michael Pollan

Introduction by Fred Kirschenmann

Designed by Roberto Carra

with Timothy Rice, Chris Blum, and Ron Bean

A Watershed Media Book
Healdsburg, California

Michael Pollan, "Don't Call it the Farm Bill; Call it the Food Bill," *The Nation*, September 11, 2006.

Fred Kirschenmann, "A Food and Farm Bill for the 21st Century."

"Fraud on the Farm," by John Burnett, National Public Radio's *Morning Edition*, November 14, 2005.

Will Allen, "The Farm Bill Cotton Bail-Out."

Laura Sayre, "Why New Zealanders Don't Like Farm Subsidies," in "Farming without Subsidies: Some Lessons from New Zealand," *New Farm*, March 2003.

Deborah Rich, "Delivering Nutrition to Seniors in Need," *San Francisco Chronicle*, June 25, 2005.

Brian DeVore, "Conservation Security Program: Rewarding the Best, Motivating the Rest."

Published by Watershed Media
451 Hudson Street, Healdsburg, California 95448
707.431.2936
www.watershedmedia.org

Distributed by University of California Press
Berkeley and Los Angeles, California
University of California Press, Ltd.
London, England
www.ucpress.edu

Library of Congress Cataloging-in-Publication Data available upon request from the publisher.

Cover Design: Roberto Carra

Printed in Canada on Rolland 100. 100 percent post-consumer, processed chlorine-free, acid-free archival paper. Color inserts printed on New Leaf Reincarnation Matte, 100 percent recycled, 50 percent post-consumer waste, archival paper.

ISBN 0-9709500-2-0
First Edition
09 08 07 06
10 9 8 7 6 5 4 3 2 1

Table of Contents

Charts & Tables

Acknowledgments

The seeds of inspiration for this book were originally sown when omnibus transportation and energy bills passed in 2005 with what appeared to be a negligent amount of public discussion, input, or cost-benefit analysis. Both legislations were padded with burdensome earmarks while lacking serious focus on conservation, climate change, regional economic development, renewable alternatives, and other bold goals so important to the challenges we face as a nation. The Farm Bill—as an economic engine driving the food and farming system—became a subject ideal for a Watershed Media project. We hope this effort helps to elevate the discussions around future Farm Bills to the status that such debates deserve and require. So many people contributed to the effort: Jennie Curtis and the Garfield Foundation, who for many years have generously offered Watershed Media their confidence and support, awarded a production grant to get this project off the ground. Nancy Shaub provided additional funding at a critical stage of the work, as did the Levinson Foundation, the Richard and Rhoda Goldman Fund, and the Laurence Levine Charitable Fund. This support is indeed humbling, and we hope our results measure up. Watershed Media was also fortunate to have many expert editors and readers who influenced the content and research: Thomas Forster of the Community Food Security Coalition; Martha Noble of the Sustainable Agriculture Coalition; Jo Ann Baumgartner of the Wild Farm Alliance; Bill Orts of the USDA Albany Research Lab; Jim Kleinschmitt of the Institute for Agriculture and Trade Policy. (That said, the final text remains solely the responsibility of Watershed Media.) Janet Reed Blake edited her third Watershed Media book. Thanks also to our outstanding graphics team: Chris Blum and Ron Bean of Chris Blum Design generated charts and graphs; Timothy Rice, our layout artist; and Roberto Carra, Watershed Media's co-founder and ever-brilliant art director. Christen and Ryan Crumley provided research and proofreading support. Many thanks to Fred Kirschenmann, Michael Pollan, Will Allen, Brian DeVore, Deborah Rich, and John Burnett for their contributions and research. Thanks also to Scott Vlaughn, Jared Lawson, and Peter Arnold, and others for their outstanding photographs. Hannah Love of University of California Press has been so helpful managing distribution arrangements for Watershed Media. We additionally would like to thank the tremendous advocacy groups and journalists who work on and cover the Farm Bill and apologize in advance if we have overlooked a contribution...

Dan Imhoff, December 2006

Mark Buell
Virginia Clarke-Laskin
Kamyar Enshayan
Karen Eggerman
Jib Ellison
Andy Fisher
Mary Grace Fowell
Sharon Wilson Géno
Randy Gray
Jon Harvey
Randy Hayes
Greg Hendrickson
Dave Henson
Quincey Tompkins Imhoff
Dana Jackson
Scotty Johnson
Diana Donlon Karlenzig
Alison Kendall
Jared Lawson
Gene Logsdon
Jim Monke
Claudia Reed
Peter Riggs
Mark Ritchie
Richard Rominger
Steve Silberstein
Jim Walters
Alice Waters
Mark Winne
Jeff Zinn

I tell you frankly that this is a new and untrod path, but I tell you with equal frankness that an unprecedented condition calls for the trial of new means to rescue agriculture. If a fair administrative trial of it is made and it does not produce the hoped-for results I shall be the first to acknowledge it and advise you.

–Franklin Delano Roosevelt, March 16, 1933,
New Means to Rescue Agriculture, speech to Congress

Men who can graft the trees and make the seed fertile and big can find no way to let the hungry people eat their produce. Men who have created new fruits in the world cannot create a system whereby their fruits may be eaten. ...The works of the roots of the vines, of the trees, must be destroyed to keep up the price, and this is the saddest, bitterest thing of all. ...A million people hungry, needing the fruit–and kerosene sprayed over the golden mountains.

There is a crime here that goes beyond denunciation. There is a sorrow here that weeping cannot symbolize. There is a failure here that topples all our success. The fertile earth, the straight tree rows, the sturdy trunks, and the ripe fruit. And children dying of pellagra must die because a profit cannot be taken from an orange. And coroners must fill in the certificates–died of malnutrition–because the food must rot, must be forced to rot.

–John Steinbeck, *The Grapes of Wrath*

The United States government's agricultural policy, or non-policy, since 1952 has merely consented to the farmers' predicament of high costs and low prices; it has never envisioned or advocated in particular the prosperity of farmers or farmland, but has only promised "cheap food" to consumers and "survival" to the "larger and more efficient" farmers who supposedly could adapt to and endure the attrition of high costs and low prices. And after each inevitable wave of farm failures and the inevitable enlargement of the destitution and degradation of the countryside, there have been the inevitable reassurances from government propagandists and university experts that American agriculture was now more efficient and that everybody would be better off in the future.

–Wendell Berry, *The Total Economy*, Citizenship Papers

Foreword

Don't Call It the "Farm Bill," Call It the "Food Bill"

Michael Pollan

Every five to seven years, the president of the United States signs an obscure piece of legislation that determines what happens on a couple hundred million acres of private property in America, what sort of food Americans eat (and how much it costs), and as a result, the health of our population. In a nation consecrated to the idea of private property and free enterprise, you would not think any piece of legislation could have such far-reaching effects, especially one about which so few of us–even the most politically aware–know anything. But in fact, the American food system is a game played according to a precise set of rules that are written by the federal government with virtually no input from any but a handful of farm-state legislators. Nothing could do more to reform America's food system, and by doing so, improve the condition of America's environment and public health, than if the rest of us were to weigh in.

The Farm Bill determines what our kids will eat for lunch in school every day. Right now, the school lunch program is designed not around the goal of children's health, but to help dispose of surplus agricultural commodities, especially cheap feedlot beef and dairy products, both high in fat.

The Farm Bill determines what our kids will eat for lunch in school every day. Right now, the school lunch program is designed not around the goal of children's health, but to help dispose of surplus agricultural commodities, especially cheap feedlot beef and dairy products, both high in fat.

The Farm Bill writes the regulatory rules governing the production of meat in this country, determining whether the meat we eat comes from sprawling, brutal, polluting factory farms and the big four meatpackers (which control 80 percent of the market), or from local farms.

Most important, the Farm Bill determines what crops the government will support–and, in turn, which kinds of foods will be plentiful and cheap. Today that means, by and large, corn and soybeans. These two crops are the building blocks of the fast food nation: a McDonald's meal (and most of the processed food in your supermarket) consists of clever arrangements of corn and soybeans–the corn providing the added sugars, the soy providing the added fat, and both providing the feed for the animals. These crop subsidies (which are designed to encourage overproduction rather than to help farmers by supporting prices) are the reason that the cheapest calories in an American supermarket are precisely the unhealthiest. An American shopping for food on a budget soon discovers that a dollar buys hundreds more calories on the snack food or soda aisle than it does in the produce section. Why? Because the Farm Bill supports the growing of corn but not the growing of fresh carrots. In the midst of an epidemic of diabetes and obesity, our government is subsidizing the production of high-fructose corn syrup.

This absurdity would not persist if more voters realized that the Farm Bill is not a parochial piece of legislation strictly concerning the interests of agribusiness farmers. Today, because so few of us realize we have a dog in this fight, our legislators feel free to leave debate over the Farm Bill to the farm states, very often trading away their votes on agricultural policy for votes on issues that matter more to their constituents. But what could matter more than the health of our children and the health of our land?

Perhaps the problem begins with the fact that this legislation is commonly called "the Farm Bill"–how many people these days even know a farmer or care about agriculture? Yet we all have a stake in eating. So perhaps that's where we should start, now that the debate over the 2007 Farm Bill is about to be joined. This time around let's call it "the Food Bill" and put our legislators on notice that we're paying attention.

Michael Pollan is the author of *The Omnivore's Dilemma: A Natural History of Four Meals*, and *The Botany of Desire*. He is a long-time contributor to the *New York Times* magazine and is the Knight Professor of Journalism at the University of California at Berkeley.

Introduction

A Food and Farm Bill for the Twenty-First Century

Fred Kirschenmann

As the manager and co-owner, with my sister, of Kirschenmann Family Farms in North Dakota, I fully understand the challenges that farmers face in today's agricultural economy, as well as the additional challenges we will likely face in coming decades. I am also keenly aware of the impact farm policy can have on farmers. The farm policies we design now will likely determine whether we will continue to have a sustainable food system in the future.

Most farmers today, even on some of the smaller and mid-sized farms, must accumulate millions in capital to acquire land and equipment just to be able to farm. And even those who successfully get into farming find most of their cash receipts eaten up each year by the expenses they incur. Consequently, most farmers stay in business only by generating additional income through off-farm jobs and government subsidies.

This economic dilemma farmers face may seem of little consequence to the average urban/suburban citizen. But an enlightened food and farm policy is of considerable consequence to every citizen on the planet.

The 1996 Farm Bill, dubbed the Freedom to Farm Act, was intended to change that. By gradually reducing government commodity subsidies and allowing farmers to produce whatever they wanted, the 1996 Farm Bill backers assumed that farmers would make adjustments in response to market demand, enabling the government to get out of the subsidy game. For example, if farmers could not generate a net profit by producing corn at the available contract price, they would presumably plant different crops. Eventually the reduced supply would drive its market price up enough for those who did plant corn to make a net profit. Of course, it didn't work out that way.

Rational farmers know that when the price of corn goes down, producing less corn to drive prices up is not a real option. They know that their individual decisions to reduce corn acres in an effort to balance supply with demand will have little effect on supply or price. It will simply reduce their own income. When the price of corn drops, they will produce as much as possible as their only defense against economic disaster. Naturally, if the price of corn goes up, they will also produce as much as possible to make up for the income lost in leaner times.

The lesson here is that, as individuals, farmers cannot manage supply to coincide with demand. That can only occur through a comprehensive farm policy. Now, this economic dilemma farmers face may seem of little consequence to the average urban/suburban citizen.

But an enlightened food and farm policy is of considerable consequence to every citizen on the planet.

It is in everyone's best interest to have:

- **Stable societies.** We cannot have stable, secure societies without a food production and distribution system that supplies a safe and adequate diet to every person. Malnutrition and starvation breed terrorism and social unrest.
- **An ecologically restorative food and fiber system.** We cannot meet present and future food needs if we continue to undermine the health of the land, pollute and overuse our water, and destroy our biological and genetic diversity.
- **An economically and ecologically efficient food and fiber system.** True efficiency must address the use of energy, capital, soil and water, and community, as well as labor. Our policies must move us toward a system based on renewable energy, recycled wastes, and diverse farming systems and ecosystems.
- **A food and fiber system that encourages independent entrepreneurship.** Human capital is critical to a sustainable food system. Without an influx of young, entrepreneurial, creative, dedicated, wise, and imaginative farmers, we will have trouble facing the challenges ahead.
- **Regional food sufficiency and food sovereignty.** We need food and farming systems that share our limited planetary resources so that citizens in every region of the planet can become food self-reliant.

The era of industrial agriculture, which relied on abundant natural resources to fuel our production systems and adequate natural "sinks" to absorb its wastes, is rapidly coming to a close. Even business design specialists now recognize this. We have so overexploited most of the earth's natural resources and so polluted the natural environment that continuing on our present industrial agriculture course is simply no longer viable.

Even oil industry leaders acknowledge that the days of "easy oil" have passed. And since our industrial food system depends almost entirely on fossil fuels, we must make major course corrections. We currently produce most of our nitrogen (fertilizer) from natural gas and most of our pesticides from petroleum; our farm equipment is largely produced with petroleum resources. And, of course, petroleum fuels all of our farm equipment.

Industrial agriculture has also depleted our fresh water resources. According to Lester Brown of the WorldWatch Institute, agricultural irrigation uses 70 percent of those resources. Four-fifths of China's grain production depends on irrigation, as do three-fifths of India's grain production and one-fifth of U.S. agriculture. And ground water resources are rapidly declining–ten feet per year in China, twenty feet per year in India.

Climate change will continue to bring us more floods, droughts, hurricanes, tornadoes, hail storms, frost, and heat waves, making it extremely difficult to maintain today's highly specialized monoculture cropping systems which require relatively stable climates.

These and other imminent challenges will force agriculture to transition from an industrial economy to an ecologically based economy. We must invent a new era of agriculture. As anthropologist Ernest Schusky has reminded us, humans have made three major transitions in the way we secure our food. First we employed hunting and gathering techniques. Then we invented agriculture and produced our food by domesticating plants and animals, using human and animal energy inputs to drive the system–the Neolithic era that lasted almost 10,000 years. In the 1930s we introduced a third era, which Schusky calls the "neocaloric era" because it depends almost entirely on imported caloric inputs–fertilizer, pesticides, antibiotics, growth hormones, feed additives, diesel fuel, and so on. We continue to use these "old calories"–that is, calories that nature has stored for billions of years–at a rapid rate, and since they are "old calories" they are not renewable. Consequently we must very soon invent the "next era" of agriculture.

This shift will require innovative planning and policy making. Can we create a new agriculture that increases the availability of healthy, nutritious food in all regions of the planet yet uses one-fourth of the external energy inputs, requires less than half the water, and thrives in adverse climate conditions?

Pulitzer Prize winner Jared Diamond vividly reminds us that those civilizations that have correctly assessed their current situations, anticipated the coming challenges, and gotten a head start in preparing for them, were the ones that survived. Those that failed in that exercise, collapsed. Shaping public food and farm policies now that begin to address these issues will help prepare us for the day when we can no longer ignore them. The critique that Dan Imhoff provides on our current farm policy in the following pages, and his invitation for all citizens to become engaged in the food and farm policy debate, can begin to take us down this new path.

Fred Kirschenmann is a Distinguished Fellow at the Leopold Center for Sustainable Agriculture and president of Kirschenmann Family Farms, a 3500-acre certified organic farm in Windsor, North Dakota. He is past president of Farm Verified Organic and has served on the U.S. Department of Agriculture's National Organic Standards Board.

Fear of ominous leap for bird flu

Fatal disease suspected of making jump among several humans for the first time

By Donald G. McNeil Jr.
NEW YORK TIMES

Reacting to the death
ay of an Indonesian m
World Health Organizat
uesday that the case app
e the first example of t
u jumping from huma
nan to human.
But the health agency
autioned that this did n
arily mean that the virus

lungs, not in the nose and throat as seasonal flu does.
The man who died Monday

d the task of finding gasolin
s things work now, every g
n or domestic — gets a tax
at subsidy is effectively ta
thanol by a tariff of 54 ce
mers see relatively little fo
nerican farm lobby is happy
his whole debate misses the
that they need alternative
forms of ethanol made from
ol to become a true alternati
ivers, Congress must set ou
e, existing subsidies.
arious bills on Capitol Hill
es, gua
age gre
c ethan
-fuel ca
It is m
using it

"Our cattle eat the trash. Little animals stick their heads in bean cans . . . until they die. There's constant harassment of wildlife."
WENDY GLENN, on immigration's effects on livestock

Wilderness victim of immigration

WHY THE FARM BILL

Divisions Remain Among EU Leaders on Farm Subsidies, Budget

European leaders at a one-day summit side London wrangled over how to accelerate their stuttering economies but failed bridge a deep gap over European Union nding or their position on global trade and other agricultural countries to modernize their economic systems in the face of globalization.

Mr. Blair had hoped to dodge the trade issue at the summit, and the divisive issue tural subsidies, wants an end to the U.K.'s €5.7 billion annual rebate.

At a postsummit news conference at Hampton Court palace outside London, Mr. Blair acknowledged that countries re- bridge the budget gap, and leaders' positions seemed frozen. Outgoing German Chancellor Gerhard Schröder warned that a new German government would not allow higher spending, saying, "Germany

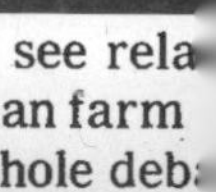

see rela
an farm
hole deb
they nee
s of etha
become a

Federal dairy advice questioned

Milk a key part of many dietary recommendations, but many can't drink it

By MIKE STOBBE
ASSOCIATED PRESS

ATLANTA — Americans should drink three cups of milk a day, the government says. Atlanta resident Kiesha Diggs ignores that advice.

Diggs, who is black, is lactose-intolerant, meaning she can't easily digest dairy products. Three cups of milk would wreak havoc on her intestines.

say information on milk alternatives is sometimes buried.
The
notch r
an adv
suit ai
ducers
with a
cause d
tose-int
Milk
the law
scaring
is not g
Many
eating e
bles, so
increas
place t

One group — which includes some nutritionists, public
s and animal
— believes di
lations should
problems and
nd them. The
ld clearly ex
groups that
eded calcium
nts like vita
ables and oth
ay.
s Joyce Guin
for an organi
mprove health
ck communi
geles area.
l understand-

California to lose out in massive farm bill

JOHN BAZEMORE / Associated

Prairie Farmers Reap Conservation's Rewards

By ELIZABETH BECKER

MEDINA, N.D., Aug. 23 — In the tawny summer landscape of the Dakota plains, where grasslands interrupt the crop sout

turn their fields into grasslands that require little upkeep and become nesting and breeding gr ducks.

The subsidies help fam

ed States Fish and Wildlife Service's habitat and population evaluation

High Price of Oil Is Biofuel Boon

Continued From First Page

spokesman for ConAgra Foods Inc., which holds a minority stake in Changing World Technologies, says the company is in a "wait and see mode" with

Transport last year installed a 5-foot-tall diesel brewer that turns cooking oil into fuel for three of the company's trucks and several forklifts. It collects its own cooking oil from local restaurants and a

loans and duction of

Ethanol: Risin oil prices boos interest in stat

to boost his investments.

The tax would aid "thos are behind the initiative," s Lundeen, a spokesman fo fornians Against Higher Ta

Khosla says his plan

acific ornia a Re- from

Report slams U.S.-European trade subsidies

Oxfam says price-boosting

of legal challenges on both sides of the Atlantic, trade and agricultural ex-

United States' corn, rice and sorghum programs, while the 25-country Euro-

of $44.8 billion, on an annual basis.

Oxfam, a nongovernmental advoca-

MATTERS

Lawmakers give nod to $100 billion farm b

Election-year measure seeks to satisfy every region, segment of country

tion on working farms.

It is also one of the major pieces of social welfare legislation before Congress this year, increasing food stamps for work-

welfare overhaul that has led to overburdened private soup

cent of the country's gross domestic product, convinced Con-

limit. The final compromise includes a limit of $360,000 but

er."

Lawmakers

ol Hill would do just that es, guaranteed loans and age greater production of c ethanol, as well as the -fuel cars, wh

e American fa

This whole

now that they her forms of e hanol to becon n drivers, Con force, existing

Various bill y providing ta rect subsidies

gallon of ethanol — x break of 51 cents. aken away for for- ents a gallon. Thus

Corn Laws for the 21st Centur

President Bush's recent proposal to suspend the tariff on imported ethanol was dead on arrival in the House of Representatives. Ethanol is an important ingredient in gasoline and an alternative fuel in

incentives away from corn-belt toward the task of finding gasoline su

As things work now, every gallo foreign or domestic — gets a tax bre bsidy is effectively taker l by a tariff of 54 cents see relatively little forei an farm lobby is happy.

hole debate misses the po they need alternative fue of ethanol made from ce ecome a true alternative , Congress must set out to sting subsidies.

s bills on Capitol Hill wou g tax incentives, guaran dies to encourage greate

Costly natural gas future predicted

Despite bulging inventories, suppliers warn no guarantee that price will remain low

By BRAD FOSS

ASSOCIATED PRESS

WASHINGTON — With the

warns there is no guarantee that the nation's supply cushion will look so comfortable by the end of summer — let alone next winter, when consumption typically peaks.

If it is an unusually hot summer, association president Skip Horvath argues, homeowners will

reason for the plunge in front-month natural gas futures since December has been weekly Energy Department data showing steady increases in the volume of gas stored in underground facilities across the lower 48 states.

On Thursday, the agency's latest report showed inventories ris-

1. Somewhere in America's Heartland...

Somewhere in America's heartland, the sun rises over a 10,000-acre cornfield. By season's end, that field, blessed with some of the world's deepest glaciated topsoil, a subterranean aquifer of ancient snow melt, and an eye-popping arsenal of John Deer equipment and petroleum products, will yield a bumper crop of 200 bushels per acre.[1]

Somewhere in Mexico, a dozen *campesinos* make their way across a desert landscape toward the 1952-mile wide border shared with the United States. For the past ten years, U.S. distributors have exported government-subsidized corn to Mexico at well-below world market prices. *Dumping* is the economic term for this, and by accepted rules of international trade, the practice is "trade distorting" and illegal. Corn farming, long a primary occupation of Central American farmers, has become a curse of poverty. If these refugees successfully navigate the perilous border crossing, survive bandits, and manage to evade or abide by *"La Migra"* (the U.S. Immigration and Naturalization Service), they may find employment in fields, orchards, vineyards, factories, restaurants, or private homes.[2] The lucky ones can then send dollars home so their families can buy the under-priced American corn that forced them north in the first place.[3]

Somewhere along the coasts of Louisiana, Mississippi, and Texas, shrimp fishermen return to port with empty nets, due to a "dead zone" in the Gulf of Mexico. This oxygen-starved, lifeless area has grown to the size of several small New England states.

Somewhere inside the vats, pipes, and tanks of a wet mill processing facility, manufacturers transform mountains of corn kernels into dozens of value-added materials. Cornstarch, corn meal, corn oil, high-fructose corn syrup (HFCS), and other additive ingredients find their way into breakfast cereals, snacks, chicken nuggets, sodas, salad dressings, and thousands of other highly processed and industrially packaged foods. Not far away, in a confinement animal feedlot operation (CAFO), 10,000 cows stand ankle-deep in their own manure as they munch away on raw corn, processed meal, and antibiotic-laced "energy and protein units" to keep pace with the world's voracious demand for meat, dairy products, and fast food. And at a gas station near you, "flex fuel" vehicle owners righteously fill their tanks with E85 ethanol fermented from surplus bushels of U.S. corn. Depending on how the numbers are crunched, however, it may require almost as much energy to synthesize liquid biofuels from cornstarch as they produce. This "net energy balance" could drastically improve with technical advances or the evolution of cellulosic ethanol, which ferments stalks, grasses, and other fibers rather than food and feed grains, but market-ready innovations are still possibly years away.[4]

In clinics and hospitals across America, physicians, nutritionists, and public health officials struggle to connect the dots of emerging epidemics: obesity, type II diabetes, coronary disease, and other dietary- and food system-related maladies. For the first time in modern history, the next generation may die younger than its parents due to dietary deficiencies. News headlines warn of surging incidences of lethal virulent diseases, such as *E. coli* infections, mad-cow disease, and avian flu, which can be transported from meat and poultry operations to humans.[5] Accounts of mass slaughtering of poultry and bans on beef exports, to contain these outbreaks, no longer shock or surprise.

Elsewhere across the country, biologists confront the realities of broad-scale pesticide and nutrient contamination of the nation's waterways.[6] This has negatively impacted fish and wildlife in a majority of our creeks, streams, rivers, and lakes. Pesticides and other farm chemicals also move up the food web, particularly affecting infants, children, women, and the elderly. Somewhere along the coasts of Louisiana, Mississippi, and Texas, shrimp fishermen return to port with empty nets, due to a "dead zone" in the Gulf of Mexico. This oxygen-starved, lifeless area has grown to the size of several small New England states. Spring runoff from the Mississippi River, loaded with nitrogen-based fertilizers from Corn Belt farms, fuels massive algae blooms that later suck the oxygen out of the water as the algae decomposes. Meanwhile in climate research stations at the extremes of the Northern and Southern Hemispheres, alarmed scientists monitor glaciers melting three times faster than computer models had projected, linking the earth's warming to twentieth-century industrial activities, including "green revolution" agriculture.

Inside the heartland's grocery chains, mothers and fathers wander the aisles, shopping for the 50 percent of meals that are still eaten and prepared inside the home. The produce sections along the periphery feature an abundance of "fresh" fruits and vegetables with an astounding tally of *food miles*.[7] Most are imported from low-cost producers out of state or out of country to provide a year-round cornucopia of berries, apples, tomatoes, salad mixes, green beans, and other seasonal crops. The interior shelves are stacked with thousands of packaged, preserved, pre-prepared, and frozen foods: most contain derivations of corn, soy, and sugar. Rather than supporting the regional economy, the lion's share of the cash residents spend on their weekly food bills leaves the region and the state. This gives rise to another modern food system phenomenon: *food deserts*.

Elsewhere in cities, rural areas, and communities across America, millions of children and adults depend on food stamp and nutrition assistance programs to fight off hunger and poverty. What money they can budget often buys foods high in calories and processed ingredients. But such food leaves them nutritionally impoverished. For school children, nutritional deficiencies can lead to life-long problems.

In the Byzantine halls of the U.S. Department of Agriculture inside the Washington D.C. Beltway, $4 billion in subsidy checks will be appropriated to help cover the production costs and incentivize this massive U.S. corn output. This record payout may only ensure, however, that farmers will break even on expenses and stay in the game to farm corn yet another season. Without direct payments, loans, and other supports that eliminate or reduce the risk in commodity agriculture, the acreage dedicated to growing corn–along with cotton, soybeans, wheat, and rice–would be far smaller. In the mean time, in communities throughout the world, similar chain reactions are set in motion as both the direct and indirect consequences of America's Farm Bill policies.

Figure 1

Effects Of Cornification

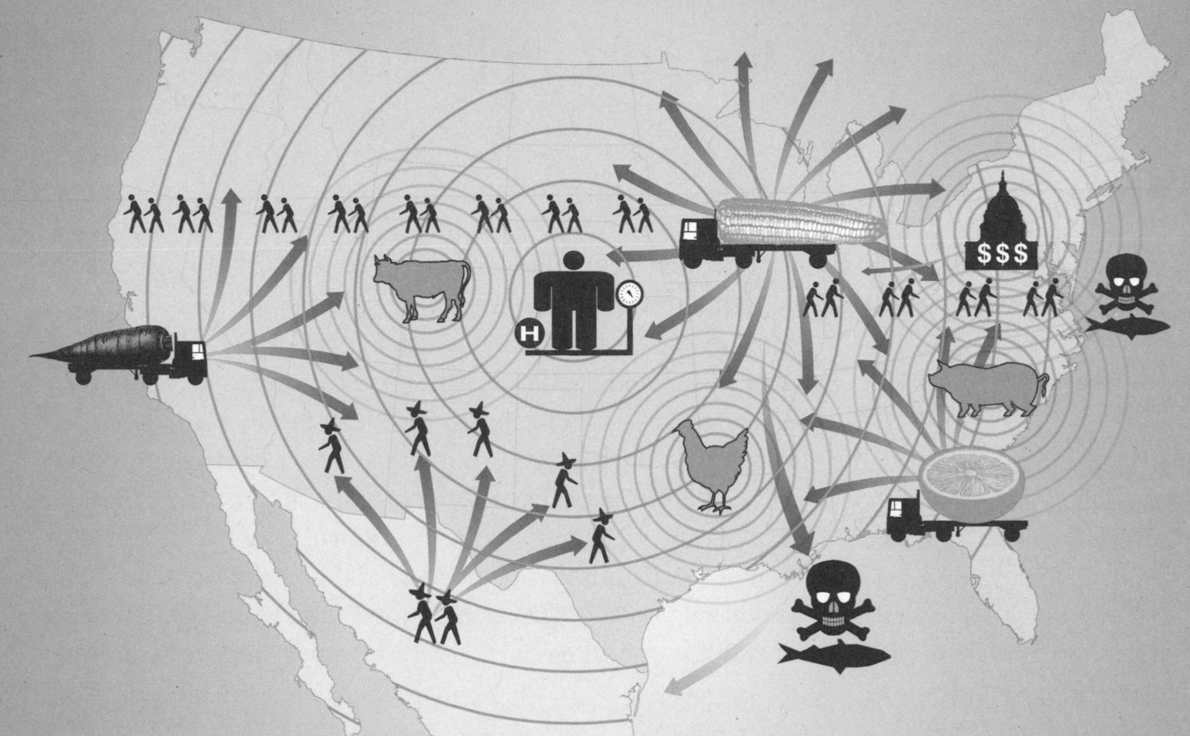

Taxpayer Subsidies. Direct payments and production incentives totaling nearly $4 billion in 2004 make corn the largest crop. Without subsidies, which affect land prices and the economic infrastructure of rural areas, farmers wouldn't plant nearly as much.

Corn surpluses. 11 billion bushels were produced on 75 million acres in 2005. Very little of the corn is actually fed to humans. Most goes to animal feed or is processed into starches, corn oil, sweeteners, or ethanol.

Rural Exodus. The farmer replacement rate has fallen below 50 percent as younger generations flee the Corn Belt for the "Fruitful Coasts." There are twice as many farmers over the age of 65 as there are under 35. Many wonder whether the United States may permanently lose the skills and productive farmland to remain an agricultural leader.

Animal Confinement Feedlots. Confinement facilities, largely made possible through the low costs of subsidized feed, house tens of thousands of hogs, chickens, or cattle. Heavy concentrations of animal wastes, odor pollution, reliance on antibiotics, and dangerous workplaces are just a few of the many concerns.

Food Deserts. Monoculture specialization on corn and other grains for export is the reason we see "so much agriculture, so little food" in farming areas. Most of the money spent on fruits and vegetables leaves the regional economy ("value leaving") giving rise to "food deserts."

The Obesity Crisis. The number of Americans who are overweight or obese climbed to 64 percent in 2006; the childhood obesity rate more than tripled between 1980 and 2004, from 5 percent to 17 percent. Lack of physical activity and poor nutrition—linked to subsidized and super-sized processed foods and soft drinks—lie at the root of the epidemic.

Food Miles. The average food item now travels approximately 1,500 miles from farm to table. California, Florida, and a number of other states (and a growing number of countries) supply the nation's supermarkets with fruits and vegetables. Relatively little of this "specialty crop" production is federally supported.

Immigration. Between 1993 and 2002, an estimated 2 million Mexican *campesinos* were forced to leave their lands and move elsewhere (the U.S. or Mexico) to attain employment. Subsidized U.S. corn, combined with the NAFTA trade agreement, have had catastrophic effects on Mexican farmers.

The Dead Zone. Nutrient and chemical runoff from farms in the Corn Belt flow through watersheds that empty into the Mississippi River and create a "dead zone" in the Gulf of Mexico, harming fish and marine life. (There are dozens of agriculturally induced hypoxic zones around the world, including the Chesapeake Bay.)

2. Why the Farm Bill Matters

If you eat, pay taxes, care about the nutritional values of school lunches, worry about the plight of biodiversity or the loss of farmland and shrinking open space, you have a personal stake in the tens of billions of dollars annually committed to agricultural and food policies. If you're concerned about escalating federal budget deficits, the fate of family farmers, a food system dominated by corporations and commodities, conditions of immigrant farm workers, the state of the country's woodlands, persistent hunger and poverty, or the marginalization of locally raised organic food and grassfed meat and dairy products, you should pay attention to the Farm Bill. There are dozens more reasons why the Farm Bill is critical to our land, our bodies, and our children's future. Some include:

What started as an ambitious temporary effort to lift millions of Americans out of the desperate economic and ecological conditions in the Great Depression and Dust Bowl, slowly devolved into a corporate boondoggle during the great family farm exodus of the 1960s, 70s, 80s, and 90s.

- The twilight of the cheap oil age and onset of unpredictable climatic conditions.
- Looming water shortages and crashing fish populations.
- Broken rural economies.
- Euphoria over agricultural expansion for the production of biofuels and bioplastics.
- Escalating medical and economic costs of child and adult obesity.
- Record payouts to corporate farms that aren't even losing money before subsidies.
- Over 35 million Americans, half of them children, who don't get enough to eat.

The Farm Bill matters because it makes some big corporations scandalously rich and drives other farmers out of rural areas–not just here, but in other countries too. The Farm Bill makes us fat and produces a vulnerable food system. The Farm Bill legalizes and supports polluting and destructive practices, then spends millions trying to put bandages on damage inflicted by past and present programs. The Farm Bill artificially sets prices and interferes with fair markets, while officials tout the virtues of "free market" and "fair trade." Its consequences include poverty, rural exodus, and famine.[8]

Although subsidies do provide a critical safety net in some years to family farms that continue to grow commodities crops, the big players and beneficiaries in the farm lobby are corporate agribusinesses, grain distributors, food processors, oil companies, absentee landlords, tractor dealers, and gasohol producers. What started as an ambitious temporary effort to lift millions of Americans out of the desperate economic and ecological conditions in the Great Depression and Dust Bowl, slowly devolved into a corporate boondoggle during the great family farm exodus of the 1960s, 70s, 80s, and 90s. As a result of the Farm Bill, citizens pay a national food bill at least three times: (1) at the checkout stand; (2) in taxes that subsidize commodity crop production; and (3) in environmental cleanup and medical costs related to the consequences of industrial commodity-based agriculture.

Most analysts, most farmers, and even many legislators agree that our present course leaves us unprepared for the urgent challenges we face in the early twenty-first century.[9] The stakes are high. And they get higher with each passing year. The silver lining is that Americans actually do have a substantially large food and farm policy program to debate. Conditions for change have perhaps never been better, as market dynamics and public awareness rapidly align to create uncertainty about farm politics as usual. Indeed, the Farm Bill matters because it can serve as the economic engine that drives small-scale entrepreneurship, on-farm research, species protection, nutritional assistance, school lunches, regional development, and habitat restoration, to name just a few. Our challenge is not to abolish government supports altogether, but to ensure that those subsidies we do choose to legislate actually serve as valuable investments in the country's future and allow us to live up to our obligations in the global community. How we get there will be a work in progress. But most observers agree that the era of massive giveaways to corporations and surplus commodity producers must yield to policies that reward stewardship, promote healthy diets, secure regional economies, and do no harm to family farms or hungry kids and their families.

Our challenge is not to abolish government supports altogether, but to ensure that those subsidies we do choose to legislate actually serve as valuable investments in the country's future and allow us to live up to our obligations in the global community.

Figure 2

Collision Course

Most analysts and even many legislators now agree that U.S. agriculture and food policy is on a collision course. It's actually heading for a "four-course collision." Urgent challenges face the food and farming system in at least four key areas: health and nutrition; economics and foreign trade; rural development and homeland security; conservation and stewardship.

For far too long, lawmakers have focused one-sidedly on narrow issues and outcomes, while critically neglecting others. The result of these political decisions is a food and farming system increasingly out of balance.

Present Challenges

Consolidation and concentration in the hands of a few corporate agribusinesses

Loss of biodiversity and wild habitats

Converging national health care crises

Childhood obesity on the rise

Chronic hunger and improper nutrition that affects nearly 40 million Americans

Sprawl into prime farmland

Record budget deficits

World Trade Organization rulings declaring U.S. export subsidies illegal

Immigration driven by "dumping" of commodities below market prices

Devastated farm economies

Rapidly aging U.S. farm population

Persistent poverty

Escalating energy costs

Increasing dependence on commodity exports and imports of "fresh" food

Water contamination and water shortages

Global warming

Increasing outbreaks of infectious diseases related to confinement livestock production

Declining honeybee and native pollinator populations

Introduction and spread of invasive species

Proposed New Directions

Caps on payments to individual recipients to level the playing field for all farmers; reforms to make feed and raw material costs more accurately align with the actual costs of production

Increase support for underfunded conservation programs

Better align crop supports with USDA nutrition guidelines

Launch nationwide farm-to-school, farm-to-college, and other fresh food distribution programs that also include a strong educational and fitness component

Maintain food assistance programs and improve access to healthy foods and adequate benefits for all Americans

Greater funding to keep farm and ranchland in agricultural use and open space rather than subdivisions and sprawl

Sharply reduce corporate giveaways and make Farm Bill spending serve as true public investment

Shift away from "trade distorting" subsidies toward green payments such as the Conservation Security Program that reward farmers for stewardship rather than overproduction of feeds

Coordinate trade and agriculture policy goals and work toward more open and fair markets

Strategic investments to revitalize and diversify rural sector

Increase support for beginning farmer programs

Honor food as a human right and foreign policy objective

Expand research into energy-effective farming systems and increase support for on-farm renewable energy infrastructure

Invest in value-added processing and flexible supports for more diversified local and regional "specialty crops"

Penalties for violators; incentives for farming systems that protect watersheds

Urgent priorities for energy conservation; increase in organic agriculture and carbon sequestration

Expansion of grass-pastured livestock operations rather than CAFOs; expansion of wetlands and wetlands protection*

Expand wild habitat for native pollinators in and around farms; adapt new programs for beekeepers

Make the spread of invasive species a research and program priority

**A United Nations Environment Program study suggests that "the loss of wetlands around the globe is forcing many wild birds onto alternative sites like farm ponds and paddy fields, bringing them into direct contact with chickens, ducks, geese, and other domesticated fowl," where the chances are greater that the bird flu virus will spread between wild and domestic birds. Wetland restoration would provide more habitat for wild birds, and reduce the likelihood they would come into contact with poultry.*

3. What Is the Farm Bill?

Every five to seven years, Congress begins the process of drafting, debating, and ultimately passing gargantuan legislation known as the Farm Bill. Each bill receives a formal name, such as the Food and Agriculture Act of 1977, the Federal Agriculture Improvement and Reform Act of 1996, or the Farm Security and Rural Investment Act of 2002, but more often people refer to each act as simply "the Farm Bill." With a history dating back to Agricultural Adjustment Act of 1933, the bill has snowballed into one of the most–if not *the most*–significant legislative measures affecting land-use policies enacted in the United States.

The Farm Bill is an *omnibus legislation* because it addresses multiple issues simultaneously. However, modern Farm Bills traditionally have two primary thrusts: (1) food stamp and nutrition programs (about 50 percent of current spending), and (2) income and price supports for commodity crops (about 35 percent of current spending). (See Figure 3, *How the USDA Spends a Tax Dollar*.) In addition to these, the Farm Bill directs and funds a wide range of other "titles." These programs include trade and foreign food aid, conservation and environment, forestry (forests and woodlots were traditionally important components of farms and should remain so), agricultural credit, rural electrification and water supply, research and education, marketing, food safety, animal health and welfare, and very recently, energy.[10] A number of policies, such as food assistance, conservation, agricultural trade, credit, rural development, and research are actually governed by a variety of separate laws, which can be, and at times are, renewed or modified as stand-alone bills. Increasingly, though, Congress finds it advantageous to combine many of these laws into a single, large-scale reauthorization of multiple statutes at the same time they renew the farm commodity programs.[11] That way, the omnibus makes us oblivious: it's nearly impossible for any single person to really understand the full extent of what the Farm Bill actually covers.

Omnibus legislation: a law that addresses multiple issues simultaneously.

Although at least half of the Farm Bill budget is targeted toward food stamp and nutrition programs that comprise an essential social safety net for tens of millions of Americans, commodity subsidies and "price supports" embody the heart of every Farm Bill.[12] At their noblest, Farm Bill payments are intended to provide stability in one of humankind's most insecure and tempestuous professions and thereby strengthen rural communities. Some programs genuinely invest in the long-term stability and stewardship of the land. (This was particularly true in the 1930s and 40s, when the bill's defining goals involved idling land in exchange for loans and price supports for storable crops.) But along the way, the Farm Bill became an engine for surplus commodity production, a gravy train for powerful corporations; and the public benefit aspects of its origins derailed.

After its radical beginnings in the Great Depression, farm and food policy developed slowly for decades with modest adjustments until political realities and global economics collided in the 1980s. Increased global trade, the call for less government spending, the concentration of distribution and processing capabilities, and low commodity prices took their toll on the farm sector and rural communities. Eventually corporate agribusinesses and mega-farms succeeded in tilting subsidies completely in their favor, but the Farm Bill has remained cloaked in a mythology that obscures its true impacts. For so long, so many things have been done in the name of the farmer rather than for the farmer that it's hard to separate rhetoric from reality.

Many Americans believe, for example, that the tens of billions of dollars the government spends on agriculture primarily support family farms. In the United States, however, three in five farmers don't get any subsidy payments at all, while the richest 5 percent average about $470,000 each.[13] Another common perception is that Farm Bill subsidies help to conserve the soil, protect clean water resources, and sustain wildlife. Yet according to the Natural Resources Conservation Service, nearly two billion tons of cropland soil is still lost every year and less than 10 percent of the USDA budget is linked to conservation practices.[14]

The most frequently heard claim is that the Farm Bill provides Americans with the cheapest and most nutritious food system in the world. But subsidized commodity-based agriculture is really designed to benefit large corporations that use and export "cheap raw materials" for livestock feed and processed ingredients. It's not exactly cheap food, either. According to researcher Charles Benbrook, if food cost is measured per calorie, rather than as a percentage of disposable income, over 20 countries enjoy cheaper food systems than the United States.[15] And the unintended costs of this heavily taxpayer-funded food and farming system are starting to mount. One-third of Americans are considered medically obese–more than 100 pounds overweight. Two-thirds of Americans are least 30 pounds over average weight.[16] Driven by government subsidized corn sweeteners, saturated fats, and food additives, this dietary crisis disproportionately affects children, people of color, and the poor.

Figure 3

How the USDA Spends a Tax Dollar

Gross Outlays Averaged Over Five Distinct Appropriations

2.7% Rural Development $2.58 billion
2.7% Research $2.54 billion
2.3% Marketing and Regulatory $2.18 billion
33.2% Commodity and Foreign Ag $31.10 billion
0.6% Administration & Misc. $0.54 billion
9.1% Conservation $8.55 billion
0.9% Food Safety $0.82 billion
48.4% Food Stamps and Nutrition $45.39 billion

Source: Congressional Research Service (data averaged over 2000, 2002, 2003, 2005, 2006).

Farm Bill Titles

The order and total number of Farm Bill titles varies from bill to bill. In the 2002 Farm Bill, the titles run as follows:

Title I — Commodity Programs
Title II — Conservation
Title III — Trade
Title IV — Nutrition Programs
Title V — Credit
Title VI — Rural Development
Title VII — Research & Related Matters
Title VIII — Forestry
Title IX — Energy
Title X — Miscellaneous (crop insurance, disaster aid, animal health and welfare, specialty crops, organic, farmers markets, civil rights, etc.)

Mandatory Spending

Commodity programs (1930s)
Food stamps (late 1960s)
Conservation (late 1980s)
Rural development & research (late 1990s)
Crop insurance (2000)

Farm Bill Names

Each Farm Bill is actually a reauthorization of the programs dating back to the 1930s as well as the authorization of new programs.

Agricultural Adjustment Act of 1933
Agricultural Adjustment Act of 1938
Agricultural Act of 1948
Agricultural Act of 1949
Agricultural Act of 1954
Agricultural Act of 1956
Food and Agricultural Act of 1965
Agricultural Act of 1970
Agricultural and Consumer Protection Act of 1973
Food and Agriculture Act of 1977
Agriculture and Food Act of 1981
Food Security Act of 1985
Food, Agriculture, Conservation, and Trade Act of 1990
Federal Agriculture Improvement and Reform Act of 1996
Farm Security and Rural Investment Act of 2002

Source: The National Agricultural Law Center

Congressional Committees Key to Farm Bill Policies

House Committee on Agriculture
Authorization committee determines the policy for the next 5- to 7-year period

Senate Committee on Agriculture, Nutrition, and Forestry
Authorization committee determines the policy for the next 5- to 7-year period

House Subcommittee on Conservation, Credit, Rural Development, and Research
Soil, water, and resource conservation; small watershed program; agricultural credit; rural development; rural electrification; farm security and family farming matters; agricultural research; education and extension services; plant pesticides; quarantine, adulteration of seeds, and insect pests; biotechnology

Subcommittee on General Farm Commodities and Risk Management
Programs and markets related to cotton, cottonseed, wheat, feed grains, soybeans, oilseeds, rice, dry beans, peas, lentils; Commodity Credit Corporation; crop insurance; commodity exchanges

Subcommittee on Specialty Crops and Foreign Agriculture Programs
Peanuts, sugar, tobacco, honey, and bees and marketing orders relating to such commodities; foreign agricultural assistance and trade promotion programs

Subcommittee on Department Operations, Oversight, Dairy, Nutrition and Forestry
Agency oversight; review and analysis; special investigations; dairy; food stamps; nutrition and consumer programs; forestry in general; forest reserves other than those created from the public domain; energy and bio-based energy production

Subcommittee on Livestock and Horticulture
Livestock, poultry, meat, seafood, and seafood products; inspection, marketing, and promotion of such commodities; aquaculture; animal welfare; grazing; fruits and vegetables; marketing and promotion orders

House Appropriations Committee
Determines how much funding should be allocated for specific programs during each fiscal year

Senate Committee on Appropriations
Determines how much funding should be allocated for specific programs during each fiscal year

Source: California Coalition for Food and Farming

4. Promises Broken: The Two Lives of Every Farm Bill...

Every Farm Bill results from at least two very distinct phases that take place over the life of the legislation. First is the authorization of the bill itself–technically the reauthorization of the existing bill dating back to the 1930s along with the introduction of any new programs. This phase lies in the hands of the Senate and House agriculture committees who negotiate a balance among the many interests served by the Farm Bill and provide directions on how taxpayer funding should be allocated.[17] This balancing is essentially a set of promises made by Congress to the public about the direction our farming and food policy will take.

Some programs acquire "mandatory funding" status, a signal that support for these programs should be made available throughout the term of the legislation. Other programs receive "discretionary funding" status, meaning their fate rests on the Farm Bill's second phase–the *appropriations* process. The final say on whether a Farm Bill program actually receives money rests with the agricultural appropriations subcommittees of the Senate and House appropriations committees. These subcommittees determine spending levels and availability of the discretionary Farm Bill programs on a yearly basis. But their powers of interpretation don't end there. The Appropriation Subcommittees can also approve yearly changes in funding to the "mandatory" programs. If Congress approves such changes through the Agricultural Appropriations legislation, the Farm Bill's funding directives for that year are overridden. (So much for those mandatory dollars promised for conservation!) *Flat funding* is the inside the Beltway term used to describe this process.

The appropriations phase means that Farm Bill funding priorities are reinterpreted every year, creating, in essence, a series of "mini Farm Bills."

As a rule of thumb, commodity price supports are the only untouchable spending categories in the appropriations process. If anything, commodity producers successfully lobby for more money, not less, through supplemental payments in response to floods, droughts, market fluctuations, or other circumstances.[18] Programs that serve the public the most, however–stewardship incentives, research funds, beginning farmer supports, farm-to-school distribution arrangements, and so on–are disproportionately the most vulnerable and normally the first on the chopping block.

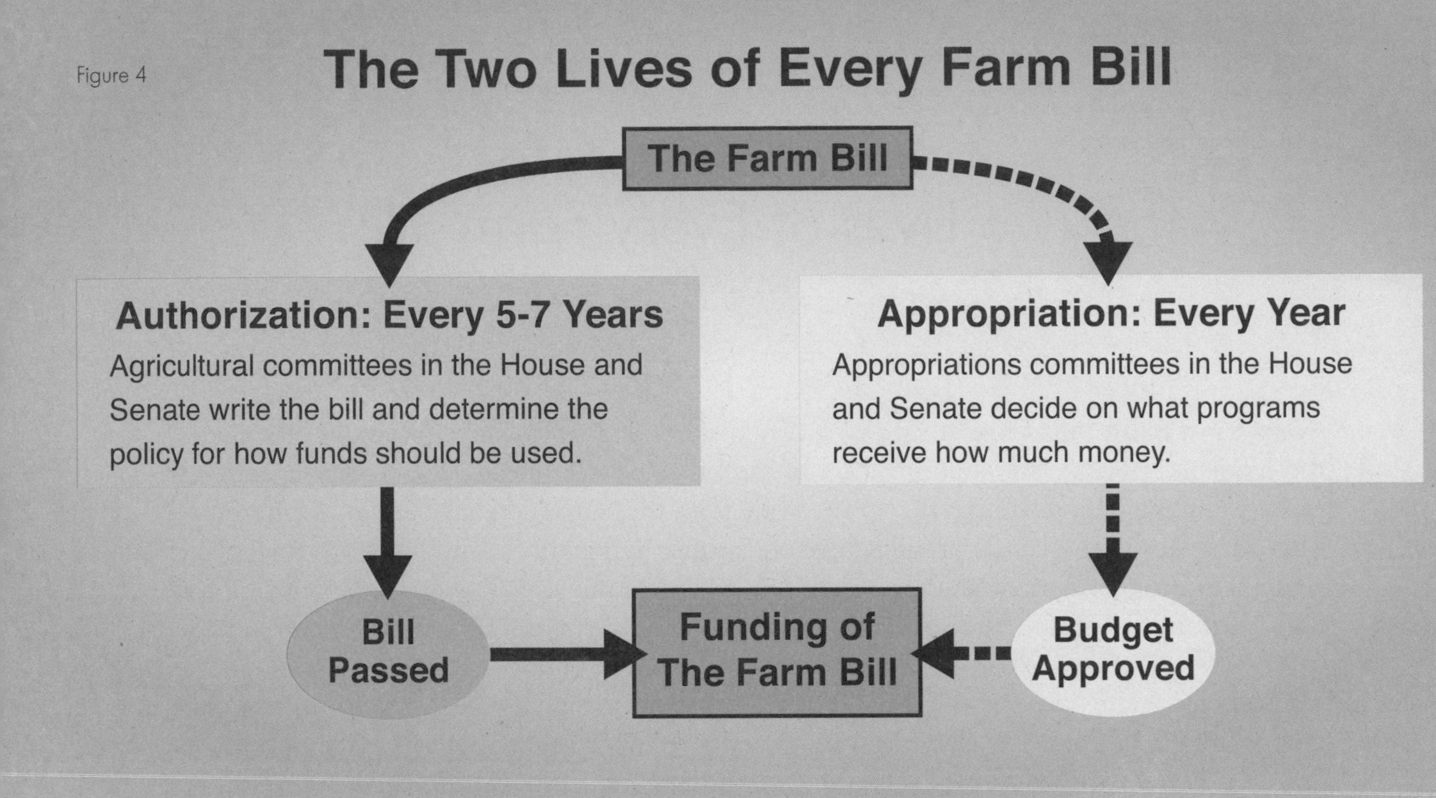

And the flat-funding doesn't necessarily end there. In response to huge budget deficits, Congress can also call for "budget reconciliation," in which the authorizing committees are directed to reopen their budgets and further decrease spending for "mandatory" programs. The end result, in essence, is a "rewriting" of the Farm Bill. In the case of the 2002 Farm Bill, the budget reconciliation process by the Senate and House agriculture committees tilted the balance among commodity programs and public interests even further, slashing conservation and farm-to-school spending, while giving away billions in loan deficiency and counter-cyclical payments to mega-farm operations.

Flat-funding: *when the moneys authorized for certain programs are cut in the appropriations process.*

The power and importance of the yearly fiscal management struggles that take place within the life cycle of a Farm Bill can't be underestimated. One recent example is the Conservation Security Program (CSP).[19] Launched as an initiative to reward sound stewardship practices rather than direct payments for maximum yields and maximum acreage, the CSP was the main concession offered to an alliance of conservationists and sustainable farming advocates during the 2002 Farm Bill negotiations. The Conservation Security Program has been widely heralded as the best way to usher in a new era of U.S. farm policies. Essentially a green payments program, the CSP meets the requirements of World Trade Organization rules on agricultural subsidies and potentially lifts the standards for farm supports by "rewarding the best and motivating the rest." Among other practices in a three-tiered stewardship program, qualifying participants must actively prevent manure from polluting waterways, limit fertilizers from entering streams, minimize or eliminate pesticide use, improve energy efficiency, and set aside habitat for wildlife. The 2002 Farm Bill promised that the Conservation Security Program would have equal funding status with the commodity title and that all farmers across the country would have access to it. Yet despite thousands of protest messages from citizens sent to agencies and representatives as of 2006, the CSP had only been funded at $489 million in a limited number of watersheds: a shortfall of 82.5 percent.[20] (See Figure 6, *Flat-Funding in Action.*) Such flat-funding is sadly typical of so many assurances to conservationists in recent years. In fiscal year 2005, highly popular Farm Bill conservation programs were cut by nearly one-third, meaning that the backlog of qualified, under-funded applications to protect habitat on both agricultural and nonagricultural lands far exceeds support.

In addition to budget slashing, appropriations committees often move beyond their fiscal mission and assume interpretive legislative powers. For example, appropriations pushed back the implementation of the "mandatory" Country of Origin Labeling program (COOL)—established to inform consumers of the origins of perishable commodities—for four years, from September 2004 to September 2008. (See Sidebar, "Totally Un-Cool," page 100.) And appropriations tinkered with organic standards by voting to allow nonorganically certified additives and ingredients in certified organic processed foods.

The *appropriations* phase means that Farm Bill funding priorities are reinterpreted every year, creating, in essence, a series of "mini Farm Bills." *Authorization* signals the beginning–not the end–to the annual *appropriation* struggles and lobbying taking place throughout every five- to seven-year journey. And if any constituency (outside of commodity producers) hopes to see the funds that were promised through the initial legislation, they must be prepared to fight tooth and nail for program dollars during annual appropriations and occasional budget reconciliation ordeals.

Figure 5

Annual Agribusiness Lobbying

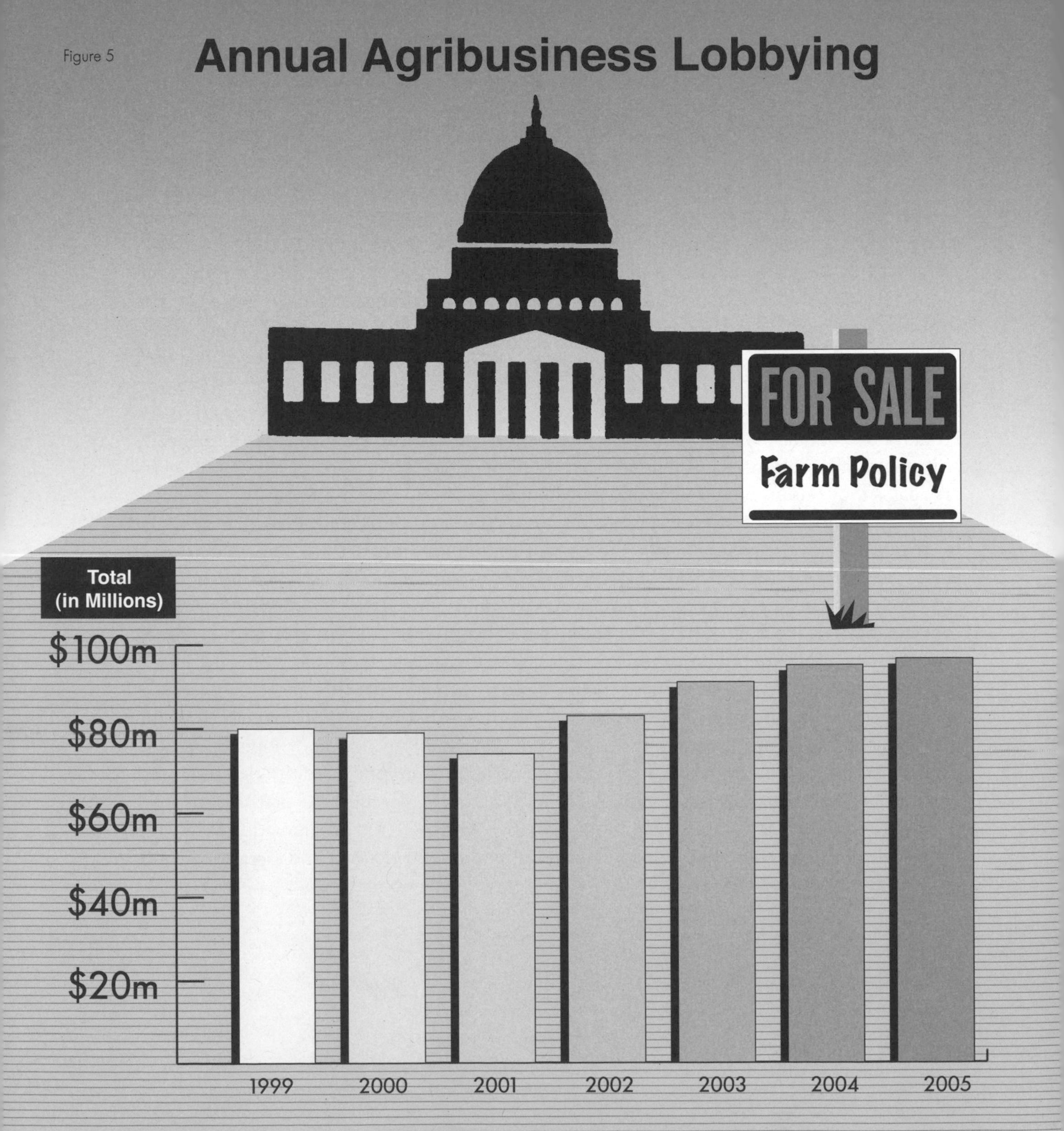

Figure 6

Flat-Funding in Action: The False Promise of Green Payments

The 2002 Farm Bill established the Conservation Security Program (CSP) as a nationwide green payments program. According to the legislation, all farmers in the country's 2,119 watersheds would be eligible to apply, with spending on equal footing with commodity payments. (Commodity payments have ranged between $15 billion and $23 billion per year since 2002.) So far, each year, CSP spending has been capped in the appropriations committees and limited to just a fraction of the country's watersheds. Ferd Hoefner at the Sustainable Agriculture Coalition reports that, at present rates, it will take 40 years to roll out the program to all farmers in all watersheds.

Year	Watersheds Appropriated	Spending
	(2,119 Total)	(Promised $3.6 to 10 Billion)
2003	0	$370 million
2004	18	$41 .1 million
2005	202	$202 million
2006	60	$259 million
2007	50	N/A

Source: Sustainable Agriculture Coalition

5. Where It All Started

The ideal of a nation built on the sweat and sacrifice of hard-working, God-fearing farmers taps a deep nerve in the American psyche. In 1801, when Thomas Jefferson became the United States' third president, 95 percent of the population made their full-time living from agriculture. Jefferson envisioned the United States as a democracy orbiting around a citizenry of yeomen farmers. He wrote:

> *Cultivators of the earth are the most valuable citizens. They are the most vigorous, the most independent, the most virtuous and they are tied to the country and wedded to its liberty and interests by the most lasting bonds. I think our governments will remain virtuous for many centuries so long as they are chiefly agricultural.*[21]

Half a century later, Abraham Lincoln planted the seeds of this vision with the establishment of the railroad land grants, the Land Grant College system, and the Homestead Act, all intended to spread independence, settlement, and stability. As the decades wore on, wave upon wave of immigrants determined to strike it rich exploited the continent's resources and natural heritage, bringing with them crops, domesticated livestock, and farming methods often not well suited to the land. By the early decades of the twentieth century, just 45 percent of the population lived on farms, and it was becoming clear that pioneering an agrarian democracy was problematic. But it took the Dust Bowl and the Great Depression to bring on total collapse.

America during the Great Depression was a hungry nation, whose most valuable natural resource—the soil—was literally blowing away in catastrophic fashion.

America during the Great Depression was a hungry nation, whose most valuable natural resource–the soil–was literally blowing away in catastrophic fashion. On a single Sunday afternoon in 1935, for example, a storm barreling through the Texas Panhandle swept 300,000 tons of topsoil into the air, choking people and animals, blanketing houses and cars, and ravaging the countryside. This one event carried twice the volume of soil excavated during the entire construction of the Panama Canal.[22]

During this time, more than a third of the U.S. population was eking out a subsistence of grinding poverty. One in four Americans still lived on farms. Increasing numbers of tenant farmers and sharecroppers were forced from their land or pushed into desperate poverty. Farm foreclosures had become commonplace. In the cities, a domino effect of bank closures and bank holidays threatened financial meltdown, while soup and bread lines grew ever longer.

By most accounts, the United States was becoming a cauldron of civil unrest. Drought, searing heat, dust storms, floods, inadequate minimum wages, maximum hours, child labor abuse and working conditions in industry, monopolistic and unfair business practices had taken a punishing toll. In *The Grapes of Wrath*, John Steinbeck described the situation this way:

> *And the dispossessed, the migrants, flowed into California, two hundred and fifty thousand, and three hundred thousand. Behind them new tractors were going on the land and the tenants were being forced off. And new waves were on the way, new waves of the dispossessed and homeless, hard, intent, and dangerous...*[23]

Ironically, the farm crisis of the 1930s was triggered not by too little food, but by too much. A decade of overzealous and speculative planting, combined with technological advances such as tractors and nitrogen fertilizers synthesized from natural gas, had yielded chronic overproduction in most crops. The "parity"—or disparity, as it were—between low farm prices and the higher costs of manufactured goods reached an ever-widening gap.[24] While low crop prices directly benefited distributors, processors, and monopolists who were increasingly dominating the food system, the U.S. agrarian culture and economy was unraveling. In order to stay afloat in a global economy, farmers planted more and more acreage. But this resulted in glutted markets, further land abuse, and prices far below the costs of production. Total farm income fell by two-thirds between 1929 and 1932. Six of every ten farms had been mortgaged to survive, and many did not. In the single year of 1932, five of every one hundred farms in Iowa were foreclosed and sold at auction.[25] Then in 1933, the price of corn actually plummeted to zero—as grain elevators simply stopped buying surplus corn altogether.[26]

In the face of such extraordinary circumstances, the Farm Bill emerged as one of the most ambitious social, cultural, and economic programs ever attempted by the U.S. government. One of the cornerstones of Franklin Roosevelt's New Deal agenda, and administered by Secretary of Agriculture Henry A. Wallace, the early Farm Bill responded directly to a number of crises:

- Rock bottom crop prices due to overproduction;
- Widespread hunger and social inequalities;
- Catastrophic erosion and soil loss due to prolonged drought and poor land stewardship practices;
- Lack of credit and insurance available to subsistence farmers;
- Need for electricity, water, and basic infrastructure in rural communities;
- Unfair export policies prohibiting free and fair trade;
- Increasing civil unrest.

Centralized food policy was not a novel concept. The Romans created a welfare system that included the distribution of free bread and grains. Their bread allotment program, which continued for centuries, was a calculated measure to stave off mob revolts. It was also only made possible by outsourcing. Rome depended almost entirely on Egyptian farmers for its wheat.[27]

Henry Wallace was a gifted, lifelong farmer, a vegetarian, and spiritual seeker, whose father had also been a Secretary of Agriculture. Under Wallace's direction, the USDA blossomed into one of the largest arms of the government, with more than 146,000 employees and a budget of more than $1 billion. (USDA Farm Bill budgets now average nearly $90 billion.) One of the driving principles of Wallace's administration was the creation of a farm support program based on a concept known as the "Ever-Normal Granary." This initiative took its historical precedent from ancient times, traceable to both Confucian China and the biblical story of Joseph.[28] The idea was straightforward though politically challenging. The government would purchase and stockpile surplus crops and livestock during good years as a protection against dwindling supply in lean times. This helped to accomplish two important goals: (1) raising market prices for farmers by contracting supply; and (2) distributing meat and grain products in times of need.

In addition, farmers participating in federally supported programs were required to idle a certain percentage of their historical base acreage, in an attempt to prevent overproduction. Author Michael Pollan describes how these price support programs worked to regulate markets:

For storable commodities such as corn, the government established a target price based on the cost of production, and whenever the market price dropped below the target, the farmer was given a choice. Instead of dumping the corn into a weak market (thereby weakening it further), the farmer could take out a loan from the government–using his crop as collateral–that allowed him to store his grain until prices recovered. At that point he sold the corn and paid back the loan; if corn prices stayed low, he could elect to keep the money he'd borrowed and, in repayment, give the government his corn, which would then go into something that came to be called, rather quaintly, the "Ever-Normal Granary."[29]

Government involvement in the food system did not begin and end with price supports and grain warehousing. Wallace's vision for farm policy included a range of departments and programs that, taken together, combined to make up an integrated food, farming, and stewardship platform. The Soil Conservation Service (originally the Soil Erosion Service and today the Natural Resources Conservation Service) addressed erosion control with alternative methods of tillage, cover cropping, crop rotation, and fertilization. Land-use planning incentives were enacted to regulate crop acreage and maximize the fallowing and recovery of fields. Programs were specifically tailored to assist sharecroppers and the rural poor. Credit and crop insurance programs provided financial mechanisms in response to the early and late season needs of farmers. Research into plant and animal diseases and new varieties and uses of crops pressed for critical innovations. Food relief and school lunch programs were part of an overall policy to provide a baseline of hunger and nutritional assistance.

Despite a demonstrated seven-times "multiplier effect" that every government dollar spent generated in the overall economy, New Deal agriculture reforms were controversial from the outset.[30] Many farmers considered hunger relief both a shameful charity and a threat to free markets. As a consequence, in the early years of the Farm Bill, millions of young hogs purchased by the government to restrict supply (bump up prices) and feed the hungry never reached their intended beneficiaries. Instead they were slaughtered and dumped in the Missouri River. Likewise, millions of gallons of milk were poured into the streets rather than nourishing famished and distended bellies.[31] Not until the term *relief* was stricken from the titles of food distribution programs and replaced with the Federal Surplus Commodities Corporation were they ultimately accepted by powerful farmer coalitions. In 1936, the Supreme Court declared that initial programs to limit acreage and set target prices for upland cotton were unconstitutional, though marketing loans and deficiency payments were later upheld.

Farmers themselves seem to have been conflicted about this emerging agricultural order and over the years have been coopted and manipulated by more powerful interests. Historian Bernard DeVito wrote that "farmers throughout the West were always demanding further government help and then furiously denouncing the government for paternalism, and trying to avoid regulation."[32] A decade prior to the 1930s Farm Bill programs, H. L. Mencken said of American farmers, "When the going is good for him he robs the rest of us up to the extreme limit of our endurance; when the going is bad, he comes up bawling for help out of the public till....There has never been a time, in good season or bad, when his hands were not itching for more."[33]

Henry Wallace, meanwhile, moved on to become vice president during the second term of Franklin Roosevelt, and his vision for farm and food policy was never integrated as he had envisioned. A genuine attempt had been made to enact policies that brought balanced abundance to the people, protected against shortages (a significant foresight considering what transpired during World War II), and buffered farmers against losses with loan and insurance programs. Ultimately, however, these programs could not solve one of agriculture's biggest challenges of the twentieth century: overproduction in a rapidly globalizing and industrializing food system.

NATION'S
FIRST WATERSHED PROJECT

This point is near the center of the 90,000 acre Coon Creek Watershed, the nation's first large-scale demonstration of soil and water conservation. The area was selected for this purpose by the U. S. Soil Conservation Service (then Soil Erosion Service) in October 1933. Technicians of the S. C. S. and the University of Wisconsin pooled their knowledge with experiences of local farm leaders to establish a pattern of land use now prevalent throughout the midwest. Planned practices in effect include improvement of woodlands, wildlife habitat and pastures, better rotations and fertilization, strip cropping, terracing, and gully and stream bank erosion control. The outcome is a tribute to the wisdom, courage and foresight of the farm families who adopted the modern methods of conservation farming illustrated here.

Erected 1955

Coon Valley Wisconsin, 1930s

This historic photo documents the collapse of the Coon Creek Watershed in Coon Valley, Wisconsin during the Dust Bowl era. The valley was also the site of the first Farm Bill conservation programs.

6. Family Farms to Mega-Farms

The overriding goals that drove agricultural policies during the critical times of the Depression and World War II–maintaining fair prices for farmers and conserving natural resources–could not be sustained. By 1949, the United States was rapidly urbanizing and suburbanizing. The 5 million farms that remained were largely homogenous: similarly sized with a fair degree of surrounding wild habitat, raising a diversity of crops, including livestock (for meat, dairy, and fertilizer), honeybees (for pollination and honey), and other products. Over a hundred commodities received some form of federal price support. All that would soon change.

The technological and industrial capacities developed during wartime were unleashed upon the civilian economy, and agriculture provided a large outlet. Tractors took on tank-like power. Chemical weapons and bombing materials were concocted into a slew of pesticides, herbicides, and synthetic fertilizers. Some of these chemicals were dropped on farm fields from the air, as squadrons of crop dusting planes joined in the Cold War effort to "feed the world." Plant breeding also evolved with Green Revolution hybrid grains tailored to these shifts in agriculture. Farming became more "capital-intensive," meaning the costs of production were on the rise. And yields continually increased, ensuring consistent overproduction and low prices for farmers.

Farmers who had maintained wild or semi-wild borders around and between fields (in accordance with the best practices of former administrations), tore out shelterbelts, windbreaks, filter strips, and contours. Wetlands were drained and forests obliterated, often with direct technical assistance and financial aid from the USDA Soil Conservation Service.

For better or worse, the Farm Bills of the New Deal era had set free the proverbial genie from the federal treasury's coffers. The temporary measures of the 1930s and 40s gradually became institutionalized. Without advocates for farmer's rights at the helm, however, the rules of the game were up for grabs, and agribusiness, along with an alliance of legislators from southeastern and upper midwestern states, learned to manipulate the system.

By the 1970s, government policy was increasingly shaped by a controversial Secretary of Agriculture named Earl Butz, who, at the end of his career, earned a reputation for uttering offensive racial and religious insults, was convicted of tax evasion, and launched a campaign to drive the final nails into the coffin of the American agrarian experience. In response to a euphoria and temporary price hikes touched off by a secret "Russian grain deal" in 1972, Butz spurred on farmers to "Get Big or Get Out," to "Adapt or Die," and to "Farm Fencerow to Fencerow." Farmers who had maintained wild or semi-wild borders around and between fields (in accordance with the best practices of former administrations), tore out shelterbelts, windbreaks, filter strips, and contours. Wetlands were drained and forests obliterated, often with direct technical assistance and financial aid from the USDA Soil Conservation Service. Yields continued to increase, and progress was measured by yields, disguising many of the effects of the industrialization of agriculture.[34]

A dark cloud moved across the agricultural landscape as this industrialization and capitalization intensified. The Ever-Normal Granary programs of the New Deal–loan-based, market and supply regulation–were quickly phased out in favor of farm crop payments based on maximizing yields. Driven by the incentive to boost acreage and crop output, farms grew bigger–even marginally productive lands were plowed up–and more specialized. Debt leveraging became business as usual. Bankruptcies and foreclosures followed, along with a rise in rural suicides and a farm exodus.[35] The American farm assumed a factory-like efficiency, and with it came the loss of natural elements such as healthy water systems, and an abundance of non-cultivated habitat where native pollinators and beneficial insects could reside. It also brought an escalation in endangered species whose habitats had been polluted or destroyed.

By the early 1980s, large grain handlers like Cargill and Archer Daniels Midland were essentially writing the Farm Bills.[36] Their mission: to ensure a steady supply of cheap commodity crops that they could trade internationally and process into value-added products.[37] (Savvy corporations set up and partook of a smorgasbord of Washington bureaucratic programs. There was money to be made in agriculture, to be sure, but not for the family farmer. For the sake of their land and lifestyle, farmers exploited themselves even more than legally permissible. And grain dealers, processors, and others willingly reaped the benefits.)

Perhaps most devastating, was that after the Butz era, the scale of agriculture had changed. With the move from family farms to mega-farms, the food and farming system reached a crossroads. Agriculture had become increasingly industrialized and dominated by concentrated corporate interests in almost every sector. Yet the sustainable agriculture movement also existed, led by small innovative organic producers driven by goals of a clean environment, healthy food, and vital communities rather than profit and market domination. In the three decades after the sustainable agriculture movement began, organic farming has become the most-consistent and fastest- growing marketing force in the global food market.

The United States now has 2,000,000 farmers but only 350,000 work full-time on farms that often require more equipment than human input. According to Peter Riggs of the Forum on Democracy and Trade, the top 150,000 mega-farms in the United States

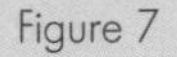
Figure 7

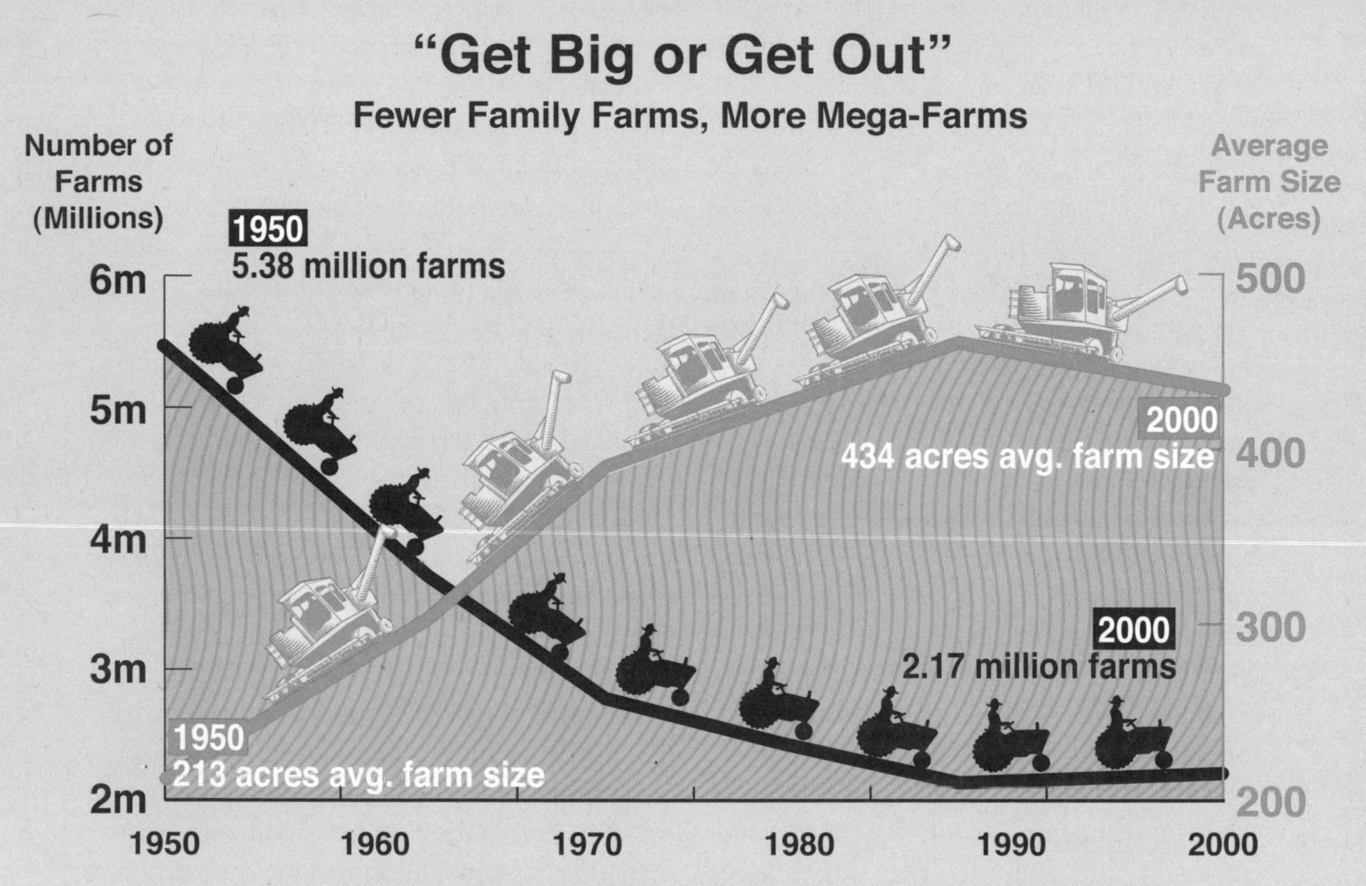

Source: USDA Economic Research Service: Structure and Finances of U.S. Farms: 2005 Family Farm Report

The "Green" Revolution

"The precolonial famines of Europe raised the question: What would happen when the planet's supply of arable land ran out? We have a clear answer. In about 1960 expansion hit its limits and the supply of unfarmed, arable lands came to an end. There was nothing left to plow. What happened was grain yields tripled.

The accepted term for this strange turn of events is the green revolution, though it would be more properly labeled the amber revolution, because it applied exclusively to grain—wheat, rice, and corn. Plant breeders tinkered with the architecture of these three grains so that they could be hypercharged with irrigation water and chemical fertilizers, especially nitrogen. This innovation meshed nicely with the increased "efficiency" of the industrialized factory-farm system…it also disrupted long-standing patterns of rural life worldwide, moving a lot of no-longer-needed people off the land."

—From Richard Manning, "The Oil We Eat"

produce most of the country's output of food and fiber. The top three or four conglomerates control at least 60 percent of the total market share in grain handling, corn exports, beef packing, and flour milling.[38] Oligopolies also dominate the crop insurance industry, the seed business, food retailing, food processing, fertilizer production, and ethanol manufacture, among many others.[39] In contrast, more than 70 percent of the rest of the world's inhabitants still make their living through agriculture, primarily at a small-scale or subsistence level.

Even conservative financial institutions recognize that commodity subsidies have led to excessive corporate concentration that is failing rural communities. The Federal Reserve Bank of Kansas, hardly a liberal think tank or family farmer–based advocacy organization, reported in 2005 that:

> *Commodity programs wed farming regions to an ongoing pattern of economic consolidation. It should not be surprising, therefore, that the very places that depend most on federal farm payments also happen to be places were economic consolidation is happening apace.... Traditional programs simply do not provide the economic lift that farming regions need going forward.*[40]

Contrary to what agribusiness might want us to believe, what's good for megafarms is usually not good for regional economies. The Federal Reserve Bank of Kansas also found that between 2000 and 2003, in nearly two-thirds of the counties that received heavy farm subsidies, the growth rate for job creation trailed the national average. A majority of heavily subsidized counties also lost population. And so it goes: Earl Butz's legacy lives on. The Farm Bill is actually succeeding at one of its decades-old policy objectives: driving small- to medium-scale commodity farmers off the land. Unfortunately, these farmers are often the most skillful, innovative, and diversified.

Big Food

Market Share Controlled by the Top Four Producers

Sector	Share	Sector	Share
Beef Packers	84%	Turkey Production	51%
Pork Packers	64%	Animal Feed Processing	34%
Pork Production	49%	Flour Milling	63%
Broiler Production	56%	Soybean Crushing	71%

Markets are considered concentrated if the share exceeds 20 percent, and very highly concentrated if the market share approaches or exceeds 50 percent.

Source: Hendrickson and Heffernan (2005); Timothy A. Wise *"Identifying the Real Winners from U.S. Agricultural Policies,"* Global Development and Environment Institute, Tufts University, December 2005.

7. The Farm Bill's Hunger Connection

The concept of a government safety net for farmers began as early as the 1920s and originally incited significant public outcry. Despite years of hard times in the rural economy, farm supports seemed to many to run against the grain of America's "pull yourself up by your bootstraps" work ethic. Even more controversial perhaps was the notion of public food distribution or financial assistance for the needy. Until 1932, that responsibility lay solely at the feet of local communities and charity organizations. Critics believed hunger relief would lead the country irreversibly toward socialism and the dole. Even as farm surpluses and global competition set off record low prices and created staggering unemployment lines during the Great Depression, no resolution seemed apparent to the paradox of want in the midst of abundance.

Having witnessed rural poverty first-hand on the presidential campaign trail, John F. Kennedy signed an executive order in the early months of his administration that revived the food stamp program.

The Federal Surplus Relief Corporation, created in 1933 as part of the first Agricultural Adjustment Administration (or early Farm Bill), was charged with purchasing, storing, and processing surplus agricultural crops to relieve the suffering caused by unemployment.[41] Although it didn't always function optimally, the connection had been made between the nutritional health of the citizenry and the nation's farm policy.

America's entry into World War II effectively eliminated agricultural surpluses, overabundant food, mass unemployment, and the need for most food assistance programs. After World War II, the New Deal surplus distribution and food stamp programs were eventually phased out, although USDA commodity food assistance lingered through the 1950s at modest levels.[42] Policy makers remained aware, however, of the previous decade's lingering effects of undernutrition and malnutrition. Throughout the World War II era, 40 percent of draftees had been rejected from duty for poor health. Hunger was no longer simply an issue of poverty or potential social unrest; a malnourished country threatened national security. With broad bipartisan and agricultural support, the federal government passed the National School Lunch Act in 1946. As its name implied, the act established school lunch programs–which included distribution of surplus commodities–throughout most of the country's public schools. It remains one of the largest and most effective public food assistance programs.

The strong economy of the postwar period, coupled with flagging political and public awareness of the problem of hunger, led to the government largely abandoning food assistance. The National School Lunch Program remained an exception. Among a few senators who picked up the torch for federal hunger and nutrition assistance later in that decade was John F. Kennedy. Having witnessed rural poverty firsthand on the presidential campaign trail, Kennedy signed an executive order in the early months of his administration that revived the food stamp program. Tensions remained high, however, between the acceptability of government food giveaways on the one hand, and the need to boost the farm economy on the other.

The initial food stamp coupons of the early 1960s were offered at a discount rather than free-of-charge. This "copayment" arrangement was intended to both dignify recipients and keep food aid a by-product of surplus crop production rather than an act of welfare. (Free surplus food was somehow politically palatable; food stamps were too much like "free money.") As the decade wore on, however, a new force emerged in Farm Bill politics. Nutrition and food assistance advocates–a.k.a. the "hunger lobby"–began to wield its power, trading votes with "farm-bloc" representatives, and ultimately gaining passage of the 1964 Food Stamp Act.

Hunger, nutrition, and civil rights advocates caught on to the potential for an expanded food stamp program to provide a much needed humanitarian safety net for a country with huge class divides. Their cause was heightened in 1968 when a groundbreaking CBS documentary, "Hunger in

Hunger in America. Among the first acts of the Kennedy administration was to reinstate the food stamp program, which eventually changed the nature of Farm Bill politics.

America," hosted by Charles Kuralt, exposed appalling conditions of poverty and malnutrition across the country. Nutritional assistance soon became an integral part of Farm Bill politics, with both annual appropriations and direct payments to families steadily increasing. By 1971, public interest lobbying organizations such as the Food Research Action Center (FRAC) and the Community Nutrition Institute (CNI) formed to defend these hard-fought gains through repeated legislative negotiations.

Food stamps have acted as a guaranteed supplement to low wages as part of every subsequent Farm Bill, assuring that most low-income Americans receive at least a minimum share in the country's agricultural output. By the mid-1970s, nearly 20 million Americans received assistance. Today the

Figure 8

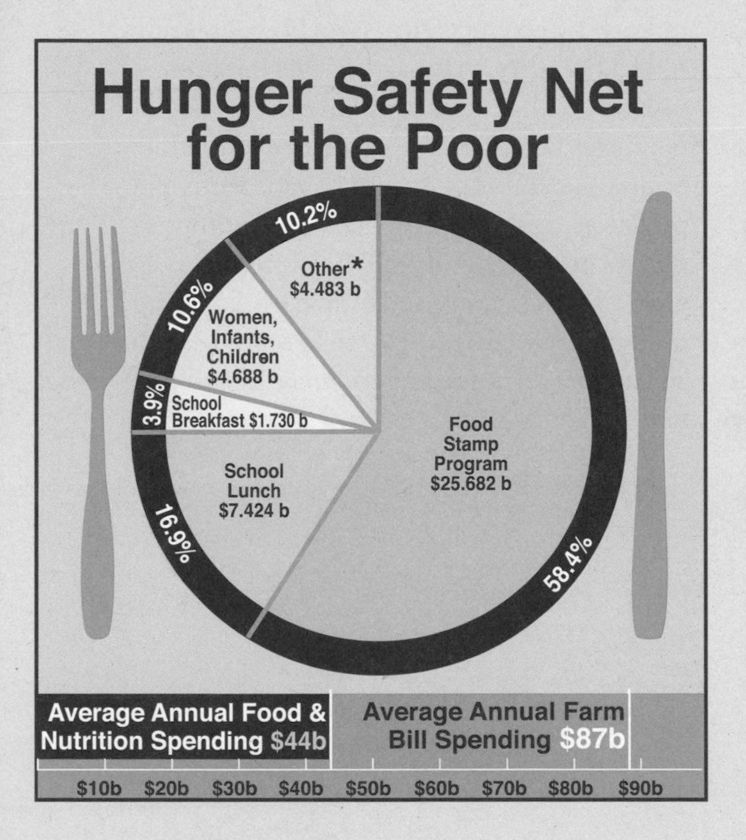

Hunger in America. The Food and Nutrition Title is by far the largest Farm Bill spending category, comprising an average of 51 percent of annual appropriations. There is a valid reason for this. The number of U.S. households classified as "Food Insecure" is experiencing significant increases every year. Food stamp participation is also on a steady rise. According to the USDA Food and Nutrition Service, in an average month of 2004, 38 million Americans were eligible to receive food stamp benefits, of which 23 million chose to participate. Typically, only 60 to 70 percent of eligible individuals actually receive food stamps.

*Other programs include: Puerto Rico Grant, Special Milk Program, Child/Adult Care Food Program, Summer Food Service Program, Child Nutrition State Administration, Commodity Supplemental, Food Distribution on Indian Reservations, NSIP Elderly Feeding, TEFAP Emergency Food Assistance, Disaster Feeding, Charitable Institutes. Source: USDA Food & Nutrition Service

Food and Nutrition Title is by far the largest appropriation of the Farm Bill, with a $60 billion budget in 2006 alone.

Until now, there has been at least one striking difference between the farm lobby and the hunger lobby, whose alliance has so influenced recent policy. Thomas Forster, of the Community Food Security Coalition in Washington D.C. explains: "Along the way, benefits to farmers got subverted as the subsidies were increasingly channeled to the very largest producers. Food policy has not been subverted from its original intent to serve as a hunger safety net for the poor."

While this may be true, a bitter irony remains. Unless the structural links to subsidized cheap food made of surplus commodity by-products are somehow radically severed, the nutritional benefits of hunger policies will remain compromised. Bellies may feel temporarily satiated, even while bodies are improperly nourished. Farm Bill programs currently don't ensure a nutritious diet for the nearly 40 million Americans eligible for food stamps today. In fact, the cheap food policy might be helping to send them to an early death according to public health and nutrition advocates.

Figure 9

Struggling to Cope

12 Percent of U.S. Households (35 million people) Were Classified as Having "Very Low Food Security" in 2005

Worried food would run out before they got money to buy more. **98%** of households

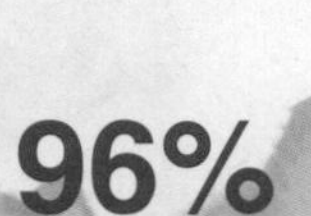

Food did not last and they did not have money to get more. **96%** of households

Could not afford to eat balanced meals. **94%** of households

Had cut the size of meals or skipped meals because they did not have enough money. **96%** of households

Had eaten less than they felt they should because they did not have enough money. **94%** of households

Had been hungry but did not eat because they could not afford enough food. **60%** of households

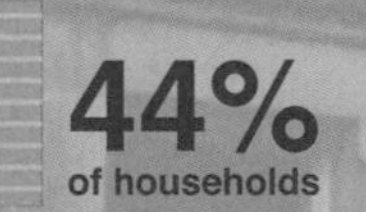

Lost weight because they did not have enough money for food. **44%** of households

An adult in family did not eat for a day because there was not enough money. **31%** of households

Food insecurity means that a household had limited or uncertain availability of food, or limited or uncertain ability to acquire acceptable foods in socially acceptable ways (i.e., without resorting to emergency food supplies, scavenging, stealing, or other unusual coping strategies).

Source: Elizabeth Williamson, "Not Enought to Eat," in "Some Americans Lack Food, but USDA Won't Call Them Hungry," *Washington Post*, November 16,2006

8. The Conservation Era Begins—Again

In 1972, a deal between international grain brokers and the Soviets launched a decade-long fury of borrowing, speculation, and agricultural expansion. Farmers in the Prairie Pothole Region, which spans parts of Iowa and Minnesota, the Dakotas, northeastern Montana, Saskatchewan, and Alberta, were caught up in the euphoria. This expanse of northern plains is known for its "potholes"–wetlands that pock the rolling hills and grasslands. These wetlands often fall within farm boundaries and serve as critical nesting and brooding habitat for waterfowl, particularly during drought cycles. With up to 70 percent of the continent's waterfowl born in this region, the Prairie Potholes are sometimes referred to as "North America's duck factory." Farmers began draining wetlands in the plow-up that the promise of new foreign grain markets inspired.

A study by the Natural Resource Conservation Service estimates an increase of nearly 26 million ducks and waterfowl in the region between 1992 and 2003, a success largely attributable to the Conservation Reserve Program (CRP).

By the early 1980s, the inevitable overproduction boom delivered a collapse of commodity prices. A prolonged drought in the Pothole region followed. Soon, the loss of vital habitat, combined with severe weather conditions, reduced the populations of ducks, pheasants, grouse, deer, and other species to record-low levels.

Legislators responded to this North American crisis within the conservation title of the 1985 Farm Bill (the Food Security Act of 1985). Funds were made available to enroll up to 37 million acres–approximately 10 percent of total U.S. farmed acreage–in the Conservation Reserve Program (CRP), essentially contracting with farmers to idle a certain set-aside acreage. That same Farm Bill included "Swamp Buster" and "Sod Buster" provisions. These "disincentive programs" immediately withdrew all federal support payments from any farmer who drained wetlands or plowed up protected grasslands. (Egregious violations of plowing up intact grasslands and plugging wetlands still occurred in many areas of the country, but it was indeed a step in the right direction to punish program abusers.)

The global grain conglomerates hotly contested these conservation programs, predicting that the protection of such large acreage would result in massive crop shortages. Throughout two decades of ever-increasing conservation program support, however, just the opposite has occurred. Surpluses persist while global commodity prices have fallen to record low levels. And each year, far more farmers are interested in enrolling acreages in conservation programs than there are funds to assist them. According to Farm Bill conservation program expert Ferd Hoefner at the Sustainable Agriculture Coalition in Washington D.C., "in 2004, three out of every four farmers and ranchers applying to participate in Farm Bill conservation programs were rejected due to lack of funds. In dollar terms, eight and a half out of every ten dollars requested were denied due to the funding shortfall. In fact, the 2004 backlog for conservation dollars exceeds the total funding available in 2005 by a three-to-one margin." Meanwhile, a study by the Natural Resource Conservation Service estimates an increase of nearly 26 million ducks and waterfowl in the Prairie Pothole region between 1992 and 2003, a success largely attributable to CRP land idling.[43]

After 1985, each successive Farm Bill added programs to the conservation title. In 1990, the Wetlands Reserve Program (WRP) provided money to set aside and restore 1,000,000 acres of wetlands.[44] It could in no way compensate for the onslaught of losses–500,000 acres per year–that had been occurring since the 1950s.[45] But it was specifically designed to identify and restore critical habitats that benefit a variety of species and protect the nation's aquatic systems.[46] Nearly 2 million acres of wetlands have been restored under the WRP, most under permanent or long-term easements. It is arguably the most successful and still among the neediest Farm Bill conservation programs.[47]

Historic Conservation and Sustainable Agriculture Program "Victories"

1985 Conservation Compliance ("Swamp Buster" and "Sod Buster" provisions); Conservation Reserve Program; National Sustainable Agriculture Information Service (ATTRA); Low-Impact Sustainable Agriculture

1990 Sustainable Agriculture Research and Education Program; Family farm, environment, and sustainable agriculture added as research goals; Integrated Farm Management Program; Wetlands Reserve Program; Water Quality Incentives Program; National Organic Program; Outreach Program for Socially Disadvantaged Farmers

1992 Beginning Farmer and Rancher Down Payment Loan Program; set-aside of loan funds for beginning farmers and ranchers

1996 Planting Flexibility; Environmental Quality Incentives Program; Farmland Protection Program; Farmers and NGO representatives added to NRCS State Technical Committees; Fund for Rural America; Community Food Grants

1998 Initiative for Future Agriculture and Food Systems

2000 Insurance Non-discrimination Policy for sustainable & organic ag; Risk Management Education Program

2002 Conservation Security Program; Conservation Partnership and Cooperation; Wetlands Reserve Program increase; Increase in Value-Added Producer Grants to pay for local, sustainable and organic marketing and processing; Beginning Farmer Credit Reforms; Organic Farming Research; Organic Certification Cost Share; Farmers Market Promotion Program; Contract Agriculture Reform; Small- and Mid-Size Farm and Rural Research

Source: Sustainable Agriculture Coalition

Figure 10

Not Enough Conservation Funds to Meet Demand

USDA Conservation Programs	FY2004 Payments ($1,000)	FY2004 Applications			FY2005 Payments ($1,000)	FY2005 Applications		
		Received	Funded	% Funded		Received	Funded	% Funded
EQIP	908,280	181,807	46,413	26%	991,879	82,114	49,406	60%
Cons. Reserve Program (CRP)	1,665,144	26,080	19,732	76%	—No general CRP signup in 2005—			
Grassland Reserve Program	69,394	10,146	1,055	10%	78,222	8,631	1,219	14%
Wetland Reserve Program (WRP)	274,769	4,219	1,035	25%	239,724	4,101	897	22%
Wildlife Habitat Incentives Program	27,828	6,045	3,012	50%	33,058	5,524	3,342	60%

Source: World Resources Institute, "Paying for Environmental Performance: Investing in Farmers and the Environment," July 2006.

Pilot programs introduced in 1996 furthered the conservation emphasis (and continue today). The Wildlife Habitat Incentive Program (WHIP) provides assistance for protecting sensitive species and restoring or maintaining critical habitats in farming regions. The Environmental Quality Incentive Program (EQIP) offers funds for a wide variety of environmental "improvements," from soil erosion and air pollution reduction, to forest replanting and thinning, to the dubious construction of expensive manure lagoons (the size of municipal sewage processing plants) that process waste in large animal confinement dairies and feedlots.[48] While landowner demand has soared for enrollment in these programs, funding remains modest at best.

Compliance versus Conservation

Taxpayers Are Footing the Bill for Feedlot Manure Lagoons

The 2002 Farm Bill conservation title spent hundreds of millions of dollars on confinement animal factory feedlots (CAFOs), not just to clean up existing pollution but also to fund new feedlots and expand old ones without accounting for their overall impacts on the environment. Under the Environmental Quality Incentives Program (EQIP) corporate feedlots are eligible for up to $450,000 (75 percent of costs) to build storage facilities for animal sewage. Massive dairies, mega–hog farms, and other confinement feeding factories are actually livestock gulags housing thousands (often tens of thousands) of animals in a single operation, producing an output of wastes equivalent to small cities. Brother David Andrews of the National Catholic Rural Life Conference has described the problem as "a fecal flood." It's an ironic—and unsavory—twist to the notion of a Farm Bill "gravy train," and a slap in the face to any taxpayer who truly cares about conservation.

Thanks to hefty campaign contributions from agribusiness lobbies and the support of a few nonprofit pollution control advocacy organizations, precious Farm Bill conservation dollars are being diverted to build and fortify manure lagoons on corporate feedlots.[49] Meanwhile, three out of every four landowners applying for help to protect and restore wetlands or idle farmlands is turned away due to a lack of funding. The issue brings to light a number of important concerns about the unwholesome interconnections between large livestock operations and Farm Bill subsidy programs.

1. **Taxpayer-funded CAFO infrastructure.** While enhancing water and air quality are indeed goals in the public's interest, should taxpayer funds be used to build the infrastructure for agribusiness to comply with regulations? Construction loans and other financing mechanisms are one thing. These cost-share programs are corporate giveaways.[50]
2. **Compliance is not conservation.** Misconstruing end-of-pipe pollution compliance as conservation is twisted

logic. Unfortunately, some politicians and even a few environmental organizations believe that the only way massive hog farms, beef, and poultry factory feedlots, and dairies will only comply with regulations is if we pay them to do so. A 2006 study by the United Nations Food and Agriculture Organization reveals that animal factory feedlots are a major contributor to climate change, generating even more greenhouse emissions than automobiles, and causing land and water degradation on a global scale.[51]

3. **CAFOs and energy production.** CAFOs are now being linked to alternative energy in a few ways. Dry grind grain-based ethanol mills depend on feedlots to consume distiller grains (protein by-products) that are not used in biofuel production. Methane digesters, which convert animal waste into combustible fuels and residual solids and liquids, are being installed on feedlots to deal with excessive manure output. Methane digesters are an extremely valuable technology, particularly for small- and medium-scale operations, but they shouldn't be used to justify nasty confinement feedlots. Proponents are calling this green power, but it is clearly more brown than clean.
4. **CAFOs and the Vegetable-Industrial Complex.** In the fall of 2006, an *E. coli* outbreak was traced to spinach farms in California's Salinas Valley. Writing in the *New York Times*, (October 15, 2006) Michael Pollan reports that "the lethal strain of *E. coli* known as 0157: H7, responsible for this latest outbreak of food poisoning, was unknown before 1982; it is believed to have evolved in the gut of feedlot cattle." Feedlot agriculture, writes Pollan, produces more than a billion tons of animal manure per year, and it often ends up in places it shouldn't, such as groundwater, irrigation systems, and industrial spinach fields. As this book goes to press, it has been unconfirmed whether the *E. coli* contamination occurred in washing facilities, through field applications of raw manure, or other avenues.
5. **End-of-pipe Band-Aids versus integrated livestock systems.** The only healthful long-term solution to the CAFO crisis is a large-scale shift away from grain-fed confinement facilities and toward perennial, grass pastured integrated livestock farms. Perhaps not surprisingly, manure lagoons have been well funded during the appropriations and budget reconciliation process of the 2002 Farm Bill. Meanwhile, the Conservation Security Program, intended to reward diversified livestock operations and other environmentally responsible farms, has seen its budget slashed (flat-funded) by nearly 80 percent. (See page 31.)

9. Freedom to Farm and the Legacy of Record Payoffs

Rhetorically, the 1996 Farm Bill—known as "Freedom to Farm"—was supposed to signal the end of the farm subsidy era. Passed by a Republican-led Congress and on the heels of a World Trade Organization agreement where developed countries committed to eliminating agriculture subsidies, legislators claimed Freedom to Farm would wean direct production supports over the following seven-year period. Among many heralded improvements was the concept of the "decoupled payment." These subsidies were no longer linked, or "coupled," to the production of a specific crop. Instead, decoupled payments rewarded landowners on the basis of their past payment history. The intent was to afford farmers flexibility to transition to new crops and alternative approaches, while the sun slowly set on the Washington subsidy game. But as the farm economy swirled into one of its cyclical tailspins, Washington reneged on the phase-out plan. Freedom to Farm triggered the largest government payouts in history, the opposite of its policy objective. Farm Bills that used to cost $3 billion to $4 billion ballooned to between $15 billion and $25 billion. Freedom to Farm also ended the era of requiring landowners to set aside acreage as a soil conservation and national supply management strategy. Instead, subsidies were linked to maximum commodity output: a fundamental shift in benefits from producers to processors and distributors.

Freedom to Farm triggered the largest government payouts in history, the opposite of its policy objective.

Even as the farm economy turned for the better, President George W. Bush inked the Farm Security and Rural Investment Act of 2002, perhaps the most lavish ever. The president boasted that this mammoth legislation "preserves the farm way of life for generations."[52] Onlookers across the country were strident. The *Washington Post* called it "a shockingly awful farm bill that will weaken the nation's economy," the *Wall Street Journal* labeled it "a 10-year, $173.5 billion bucket of slop," while the *Greensboro News Record* deemed it "a gravy-train for mega farms and corporations."[53]

This time the 2002 bill introduced "counter-cyclical payments" that fluctuate depending on global market prices in an attempt to do away with a recent rash of blockbuster disaster bailouts. It also set a record for conservation spending (theoretically at least), including the Conservation Security Program (CSP), which holds the potential to revolutionize

Figure 11

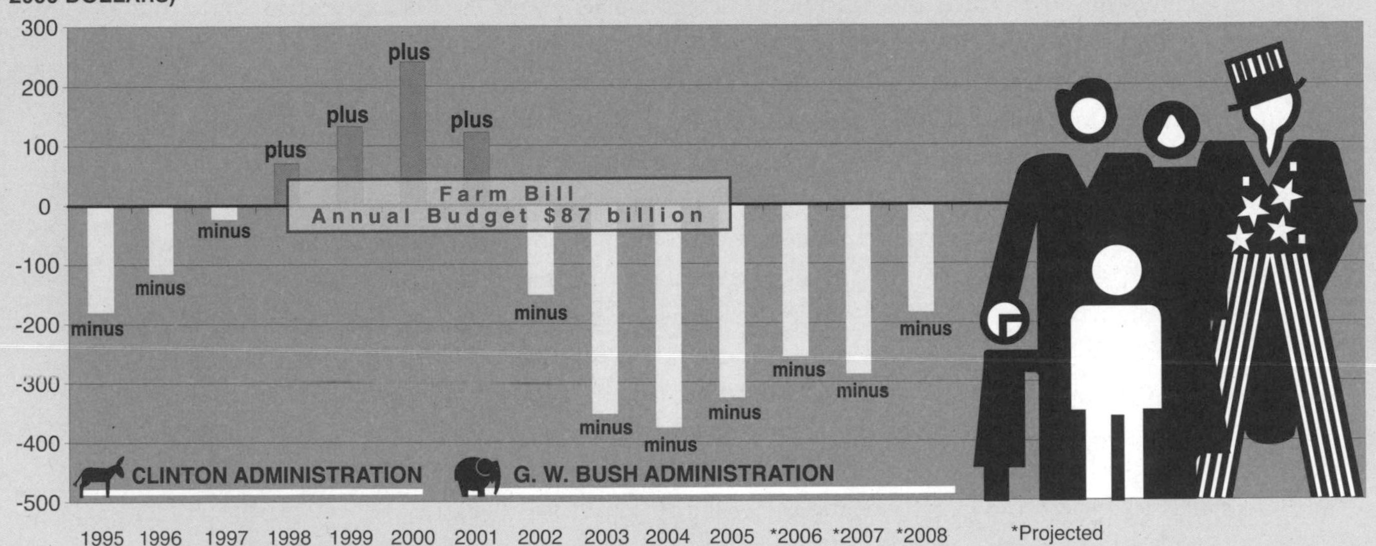

Source: Historical Tables, Budget of the United States Government, Fiscal Year 2007; Congressional Budget Office.

Budget deficits will challenge spending decisions. The mounting costs of the prolonged war in Iraq, Medicare, Social Security, budget deficits, ongoing trade imbalances , and unintended national disasters like Hurricane Katrina will force legislators to scrutinize all spending. The hard and honest truth is that—with the exception of record payouts for commodity producers—many programs have already been cut to the bone through the annual appropriations process. Unfortunately, the price of doing nothing to address the complex interrelated challenges of current food and farm policy is unaffordable.

agricultural subsidies, and the Grassland Reserve Program (GRP), intended to protect rare remnant prairies and grassland habitats. But conservation programs in general have been drastically underfunded, with as many as four out of five applicants turned down for programs due to lack of support. The overall results have been all too predictable. With so little money dedicated to sound stewardship and diversified farming, the only logical incentive is to "farm the system" by producing as much as possible. The largest and most aggressive operators receive the most benefits and use these subsidies to drive up cash rents and land prices, forcing small and medium-sized farms out of business. Soil conservation and soil building, crop diversity and crop rotations, watershed protection and water conservation are no longer values, virtues, or best management practices as they had once been in an era of much smaller farms. In 2005, alone, when pretax farm (agribusiness) profits were at a near-record $72 billion, the federal government handed out more than $25 billion in "aid" to commodity producers, almost 50 percent more than it paid to families receiving welfare.[54]

Despite billions of dollars devoted to conservation programs, the assault on remaining intact grasslands and wetlands continues. Ducks Unlimited estimates that 22 million acres of grasslands in the Prairie Pothole Region, which are essential to sustaining North American waterfowl populations, remain at risk to the plow. Subsidies are again the culprit. Countercyclical payments, crop insurance, disaster relief, and other Farm Bill price supports have virtually eliminated the risks for already wealthy commodity agriculture operators. Ducks Unlimited fears an aggressive expansion into important remaining native habitats that were previously uneconomical or impractical to farm. Only a renewal of Conservation Reserve Program contracts[55] as well as true enforcement of a "Sod Buster" or "Sod Saver" mechanism that punishes landowners for plowing critical habitat and a significant increase in conservation measures, they argue, can directly influence land stewardship and help keep healthy wildlife populations in the air.

Fraud on the Farm: Tomato Farmers Caught Out in Insurance Scam

John Burnett

Morning Edition, November 14, 2005— A wheatfield in Georgia, a soybean field in Iowa, a cotton crop in West Texas, a tomato farm in Tennessee: These are crime scenes.

The felons are a small group of farmers who falsely claim that weather ruined their crops so they can collect the insurance. The U.S. Department of Agriculture says they cheated the U.S. Treasury and insurance companies out of $160 million last year. An NPR investigation reveals this crime is growing in size and complexity, while some insurance companies look the other way.

Interviews with 50 individuals in eight states—investigators, prosecutors, farmers, watchdogs, and government regulators—reveal a culture of cheating that has grown up among a small group of farmers who exploit the nation's government-backed crop insurance program. The program was tailor-made for fraud.

To understand how it's done, come down to the fertile valleys on the border of North Carolina and Tennessee, where the federal government has just wrapped up the largest case of crop insurance fraud ever uncovered.

According to Gretchen Shappert, U.S. attorney for the western district of North Carolina, "the Warren Farms investigation is literally the mother of all crop fraud investigations. It was a result of a perfect storm of individuals who were involved in fraud."

Robert and Vicki Warren are among eight people who pleaded guilty to swindling the government and insurance companies out of more than $9 million in bogus insurance claims from 1997 to 2003. The Warrens were among the largest tomato growers east of the Mississippi; at one point they owned 26 farms in three states, including one run by Bobby Chambers.

"We grow different kinds of produce, tomatoes, green beans, cucumbers, just a little of everything," says Chambers, a beefy, baby-faced, 42-year-old lifelong farmer who runs a spread that borders the Nolichucky River in Cocke County, Tennessee.

According to trial records, he helped the Warrens stage a hailstorm to make it look like their tomatoes had been destroyed, so they could collect the insurance money.

Chambers says he bought a bag of cocktail ice and a disposable camera, and, on his boss's order, created a foul-weather tableau. "The way we did it, we was down taking pictures, out this row, and then we just stood behind it and threw the ice over the top. To me, it looked like a hailstorm," says Chambers.

To complete the scene of devastation, they then picked up wooden tomato stakes and attacked the unsuspecting vegetables. "They had one Mexican who did all the beating, he beat every 16,000 of them. He'd just go through there and knock the leaves off of them," says Chambers, as he illustrates the activity with a long stick. "It made it look like the hail had beat it up."

To understand the crime, you've got to know who the players are.

- The farmer buys an insurance policy that provides partial coverage—usually 50 to 60 percent—for the crop he expects to raise.
- The insurance agent sells him the policy.
- The loss adjustor is dispatched to inspect the field if the farmer claims a disaster.
- If the disaster is confirmed, the crop insurance company sends the farmer a check.

And the U.S. Treasury, which guarantees the riskiest farm insurance, often reimburses the insurance company.

And there's one more player: the federal prosecutor. In this case, Richard Edwards was the prosecutor who blew the whistle on the farmers, the agent, and the adjustor.

"The thing that was so interesting to us about this case was all of the classic methods—and the mind of man knows no limits in ingenuity to commit crime—seemed to show up in one case," Edwards says.

Edwards, along with investigators with the USDA, spent three years unraveling and prosecuting the Warren case. They found tens of thousands of falsified documents: false planting dates, false farming histories, overstated acreage, fake shipping manifests, yield records shifted between fields, hidden production, front producers, as well as the photographs of the cocktail-ice hailstorm.

The fake weather disaster was poorly executed. There was scant evidence 1.5 million tomato plants had ever been planted, as the Warrens claimed. But Edwards says when the loss adjustor came out he approved it anyway.

"Bobby Chambers's testimony was that it was a 'drive-by adjusting,' that they were only there a few minutes, and looked like they didn't want to get their shoes dirty."

Even the USDA was unwittingly complicit in the scheme.

The government is so generous with crop insurance that it subsidizes farmers' premiums. Edwards says the USDA paid the Warrens more than $2 million to help them insure their tomatoes. He compares it to the following hypothetical situation: "Every year a bank gets robbed and they notice the bank robber is using an old getaway car and they ask, 'Would you like a car loan to have a nicer getaway car next year when you come to rob us?' Because the government is subsidizing the farmer's ability to defraud us for the coming season."

Bobby Chambers testified for the government and got two years probation for his role in the scheme. The adjustor and the insurance agent have been convicted, but not yet sentenced. Robert and Vicki Warren each received some six years in federal prison.

Sean Devereaux, the attorney for Robert Warren, says that while his client admitted guilt, it's the whole system that's corrupt. "It's fine for the government to issue sentencing memoranda and make Robert Warren appear to be the Saddam Hussein of crop insurance, but he's not," Devereaux says.

"He basically was approached by people selling insurance and told, 'This is an easy thing to do. Don't worry, this is the government's money, it's not the insurance company's money.'"

The federal indictment, in fact, states that the Virginia-based insurance agent coached the Warrens, in detail, about how to perpetrate the fraud. What's more, the adjustor testified that his supervisor at the insurance company—Fireman's Fund Agri-Business, one of the largest in the country—instructed him to lie on crop-damage forms for the Warrens.

A spokesman for Fireman's Fund said neither the adjustor nor his supervisor work for the company anymore.

According to Michael Hand, who is in charge of compliance at the Risk Management Agency, the federal agency that administers crop insurance, "Anytime you have cases like the Warrens, that's something that was out of control and never should have happened...Unfortunately, I have a number of cases like that, active and settled, at this point, that exceed a million dollars."

Prosecutor Edwards didn't know anything about the obscure world of crop insurance before he worked on the Warren case. Now he's something of an expert, and he says he's surprised at how easy it is to flimflam the program. "The American taxpayer is getting defrauded out of millions and millions and millions of dollars. The Warrens are in no way unique," he says.

Everyone interviewed for this series agreed that the overwhelming majority of the 788,000 American farmers who buy crop insurance are honest. USDA officials estimate about 5 percent of indemnities paid out each year go to phony claims, about the same proportion found in other types of insurance.

What's different in this case is that crooks are increasingly brazen, according to the USDA Inspector General's Office; the money they're stealing usually comes out of the U.S. Treasury—and powerful political forces have resisted reform.

John Burnett is a journalist for National Public Radio who lives in Austin, Texas.

10. Who Gets the Money?

Who gets the money? For the simplest answer, one might swipe a line from the 42nd president of the United States, Bill Clinton. "It's the commodity groups, stupid." Of the $112 billion U.S. taxpayer dollars spent on commodity subsidies between 1995 and 2004, more than 80 percent went to the production of just five crops: *corn, cotton, wheat, rice,* and *soybeans.* Half of that money went to the seven states that produce most of those commodities. The richest 10 percent of farm-subsidy recipients (many of whom are corporations and absentee landowners and can hardly be classified as growers) took in two-thirds of those payments. For many of the country's 2 million remaining farmers who receive no payments at all and survive primarily on off-farm income, equating the Farm Bill with "saving the family farm" adds insult to injury.

Out of the hundreds and even thousands of plant and animal species cultivated for human use, the Farm Bill favors just four primary groups: food grains, feed grains, oilseeds, and upland cotton. Most of these are either fed to cattle in confinement or processed into oils, flours, starches, sugars, or other industrial food additives.

Very little of all that subsidized output is edible, at least by humans. Out of the hundreds and even thousands of plant and animal species cultivated for human use, the Farm Bill favors just four primary groups: food grains, feed grains, oilseeds, and upland cotton.[56] Most of these are either fed to cattle in confinement or processed into oils, flours, starches, sugars, or other industrial food additives. While the USDA's Food Pyramid emphasizes the nutritional advantages of eating five daily servings of fruits and vegetables, financial support for diversified row crop and orchard farming remains relatively disconnected from the balanced, healthy diet that professional nutritionists endorse.

Farm subsidies discriminate geographically as well. With 2,000 miles of waterways, nearly 30,000 farms, and over $30 billion in annual on-farm revenues, California tops all states in terms of sales, yet 90 percent of its growers receive no agricultural subsidies. (California contributes more than 12.5 percent of the total U.S. agricultural market value and nearly half of all fruits, nuts, and vegetables, yet its farmers receive less than 5 percent of commodity payments). The state's relatively small number of cotton and rice farmers collect two-thirds of all in-state payments.[57] Some, like the J. G. Boswell Company, one of the country's largest farming operations, receive scandalously high amounts. J. G. Boswell routinely cashes seven-digit subsidy checks for growing cotton–a surplus, water- and chemical-intensive crop–in the bottom of what was once the largest freshwater lake west

Figure 12

Following

Congressional District Distribution of Farm Bill Dollars Subsidy Payments 1995–2004

<$300 million	. . . .	344 districts, 15% of payments
$300 million–$800 million	. . . .	44 districts, 15% of payments
$800 million - $1.66 billion	. . .	25 districts, 20% of payments
Over $1.66 billion	. . . .	22 districts, 50% of payments

Sources: (1) Congressional District Mapping Information, Environmental Working Group. (2) Top 20 Subsidy Recipients 1994–2004, Environmental Working Group (3) Commodity Group breakdown, Congressional Research Service.

Too Few Eggs in Too Few Baskets?

84% of Commodity Credit Corporation spending is directed to just five crops:

Corn 32% Cotton 20% Wheat 13% Soybeans 13% Rice 6%

the Money

Top 20 Subsidy Recipients 1994–2004*

Recipient	Location	Amount
Riceland Foods	Stuttgart, AR	$ 533,757,334
Producers Rice Mill	Stuttgart, AR	$ 295,006,023
Farmers Rice Coop	Sacramento, CA	$ 143,546,751
Harvest States Cooperatives	Inner Grove Heights, MN	$ 44,422,241
Tyler Farms	Helena, AR	$ 37,010,598
DNRC Trust Land Management	Helena, MT	$ 32,049,365
SD Building Authority	Sioux Falls, SD	$ 28,193,801
Pilgrim's Pride Corporation	Broadway, VA	$ 26,461,206
Ducks Unlimited	Jackson, MS	$ 25,648,392
Missouri Delta Farms	Sikeston, MO	$ 25,360,161
Montana BD Investment	Billings, MT	$ 20,197,864
Cargill Turkey Products	Harrisonburg, VA	$ 17, 593,150
JG Boswell Company	Corcoran, CA	$ 17,290,870
Bia	Ada, OK	$ 17,055,635
Dublin Farms	Corcoran, CA	$ 13,543,970
Morgan Farms	Cleveland, MO	$ 13,379,656
Due West	Glendora, MS	$ 13,233,613
Napi	Farmington, MN	$ 12,643,037
Colorado River Indian Tribes Farm	Parker, AZ	$ 12,277,405
Walker Place	Danville, IL	$ 11,823,765

*Recipients could include multiple growers, entities, or cooperatives.

Subsidy Sleuthing

Thanks to the work of a growing number of governmental and nongovernmental resources, following the Farm Bill subsidy trail is not that difficult. The Environmental Working Group's "Farm Subsidy Database" (see www.ewg.org) provides extensive listings, analysis, and regular updates. Newsletters such as the Sustainable Agriculture Coalition's "Making Hay" (www.sac.org), Defenders of Wildlife's "Rural Updates" (www.familyfarmer.org), and the Center for Rural Affairs' "National Rural Action Network" (www.cfra.org) are focused on the issue. Oxfam International (www.oxfam.org) monitors subsidies internationally. Economists at the Congressional Research Service and USDA Economic Research Service (www.ers.usda.gov) offer extensive reports on Farm Bill authorization and appropriations.

Figure 13

of the Mississippi. Just 40 percent of the U.S. cotton crop is used domestically. The rest is dumped on the world market at lower than market prices, impacting poorer but often more efficient farmers.[58]

Florida is another prolific food producer, with extensive citrus, row crop, dairy, and calf breeding operations. Yet only 6 percent of Florida's farms and ranches receive direct subsidies. According to the Environmental Working Group, if farm payments were based on overall contributions to the nation's food and fiber supplies rather than the narrowly targeted commodity groups, five other states with large, but mostly unsupported, farm sectors would immediately benefit: North Carolina, Pennsylvania, Washington, Oregon, and Colorado.[59] Other regions traditionally left out would also get their due on par with their food and fiber output: all of New England (from Maine to New Jersey); the mid-Atlantic (from Georgia to Maryland); most of the upper Midwest; and states scattered across the South and West.[60]

Scotty Pippin, The Prince of Lichtenstein, and Christmas Trees

Another troubling aspect of the money trail is the lack of practical limits on how much a single farming operation can receive. Thanks to numerous legal loopholes, lax enforcement, and loose definitions of "engaged in farming," essentially no spending caps exist in the current subsidy system. This is not for a shortage of public outrage or congressional grandstanding, however.

Since the 1970s, federal law has capped subsidy payments at $50,000 per year. But farmers have easily circumvented this by morphing into numerous entities (sometimes referred to as Christmas trees) that each become eligible for payouts. Then in 1986 a nationwide scandal erupted when reports surfaced that the Prince of Liechtenstein had collected more than $2 million in cotton and rice subsidies as an absentee landowner.[61] In response, Congress created the so-called three-entity rule. Under this provision, a farmer could collect $50,000 in subsidies in his own name, and, as half-owner, up to $25,000 for each of two other entities.

But since the 1987 limits were enacted, new subsidy programs have proliferated and a succession of new loopholes have further eroded subsidy caps. Farmers now form complex family partnerships with associated limited liability companies that branch new tentacles into the subsidy gravy train, stretching the legal definitions of what it means to be "actively engaged in farming." Lawyers and accountants opportunistically pander to these loopholes, offering "payments limitations planning" services that turn family farms into hydra-like corporate subsidy schemes. According to a recent study in the *Atlanta Journal-Constitution*, for example, in 2005 at least 195 U.S. farming operations collected more than $1 million each from taxpayers.[62]

European princes haven't been the only targets of citizen indignation over subsidy eligibility. The "Scotty Pippin Rule"–named after the multimillionaire NBA basketball player whose farm subsidy receipts made headlines–determined that noone with an adjusted gross income over $2.5 million of which less than 75 percent came from farming, could receive program supports. But rules are one thing. Enforcing those rules has become a bureaucratic nightmare. In 2004, the Government Accountability Office (GAO) conducted a study and found that at least half of the time USDA Farm Service Agency field offices failed to use their own tools to determine eligibility.[63]

With failed payment limit regulations, a lot of non-farmers receive subsidy payments. According to a 2006 report by the *Washington Post*, the federal government paid $1.3 billion between 2000 and 2006 in rice and other crop subsidies to landowners that did no farming at all.[64] Included in this group were subdivision developers who bought farmland and advertised that prospective homeowners could collect subsidies on their new backyards.

Subsidy Tracking

In October 2006, *Atlanta Journal-Constitution* reporters Ken Foskett, Dan Chapman, and Megan Clarke published a four-part investigative series on farm subsidies based on interviews conducted with more than 200 farmers, economists, and government officials, and on databases covering 12 years of USDA payments.

Here's what they found out:

- Programs reward the largest producers, regardless of need, income or exposure to risk. Nearly $10.5 billion—almost 50 percent of all commodity subsidies—went to 5 percent of eligible farmers in 2005.[1]
- In Georgia, just 4 percent of eligible farmers receive half of all subsidies. Those beneficiaries averaged $200,900 in subsidies in 2005. Average for the other 96 percent: $8,300.[2]
- Nationwide at least 195 farming operations collected more than $1 million each from taxpayers in 2005.[3]
- Subsidies help the largest farms to acquire the best land and squeeze out smaller growers. Georgia cropland prices rose by 65 percent in just the past two years, at least in part because of guaranteed government payments.[4]
- The USDA pays farmland owners regardless of whether they farm the land themselves. Hundreds of millions of dollars have been paid to urbanites and large institutions, including major universities and Fortune 500 companies.[5]
- Only one farm in four produces a commodity eligible for a subsidy, primarily cotton, rice, corn, wheat, or soybeans.[6]
- From 2000 to 2003, the growth rate for jobs trailed the national average in nearly two-thirds of counties getting heavy subsidies according to the Federal Reserve Bank of Kansas City, Missouri. A majority of heavily-subsidized counties lost population.[7]
- Commodity certificates totaled $5.7 billion between 1999 and 2006. Government marketing loan programs pay the difference between the amount of the loan and the market value of a farmer's crop, up to $75,000. Commodity certificates allow a farmer to keep amounts over the $75,000 limit.[8]
- Countercyclical payments, with a limit of $65,000, cost $7.7 billion between 2002 and 2006. These payments kick in when a crop price falls below the target price set by Congress.[9]

Notes 1-7: Ken Foskett, Dan Chapman, and Megan Clarke, "How your tax dollars prop up big growers and squeeze the little guy," *Atlanta Journal-Constitution*, Oct. 1, 2006.; **Notes 8-9**: Ken Foskett, Dan Chapman, and Megan Clarke, "How savvy growers can double, or triple, subsidy dollars," *Atlanta Journal-Constitution*, Oct. 2, 2006.

11. The Multiple Benefits of Skilled Farmers and Healthy Rural Lands

As lopsided and dysfunctional as the subsidy system has become, we are still fortunate to have a Farm Bill to debate. Certainly taxpayer-funded support for agriculture should be predicated on a social contract or *quid pro quo*, if you will: *providing a financial safety net for farmers and the hungry and investment in rural America in return for a secure, healthy food system and the long-term benefits of land stewardship.* What could be more important to any country's interests than maintaining a diverse crop base and production capacity, preserving clean water sources and open space, protecting wildlife habitat, and enhancing a culture and economy that emphasizes regional food security and the conservation of agricultural resources? But how do we quantify and qualify these benefits? And how do we spend to benefit the maximum number of farmers and citizens alike?

"If we are not careful," secretary of the California Department of Food and Agriculture, A.G. Kawamura warned at a public forum in 2005, "we could lose the farm and the food system on our watch." Kawamura was no doubt alluding to the possibility that agriculture might become a totally outsourced activity–even less important economically than it is today. (Agriculture now represents just 2 percent of Gross Domestic Product; Services comprise 70 percent and Manufacturing 27 percent of GDP.) Decades of consolidations in farming and agriculture have resulted in a steady decline in rural areas. Youth continue to flock from the corn-rich "heartland" to what is sometimes referred to as the "fruitful coasts," where farmers produce a majority of the country's fruits and vegetables. Younger Americans are simply unable to afford or are increasingly unwilling to accept the economic hardships of farming. Meanwhile, each year the United States edges toward becoming a net importer of foods. (See Figure 14, *Who Will Grow Our Food?*) Farmers over 65 years of age now outnumber those under 35 by a ratio of more than two to one. Over the next two decades, according to the Food and Farm Policy Project, 400 million acres of agricultural lands–roughly five times the amount of land set aside in our national parks–will be transferred to new owners. Whether this acreage continues

Farmers over 65 years of age now outnumber those under 35 by a ratio of more than two to one. Over the next two decades, 400 million acres of agricultural lands–roughly five times the amount of land set aside in our national parks–will be transferred to new owners.

as productive farmland, open space, forest, or rangeland will largely depend on the incentives, education, and infrastructure luring younger generations to continue with this most basic profession.

The economics of agriculture have always been extremely challenging as farming rarely conforms to the high-return expectations of Wall Street. While local costs in agriculture vary from place to place, prices are determined globally. Growing conditions are subject to the vagaries of weather. Farmers have been caught in cycles of overproduction for more than a century, increasing their efficiency and expanding acreage, only to earn ever-marginal returns in bloated markets. For these and other reasons, leading economists such as John Maynard Keynes have long argued that subsidies can serve as essential fiscal tools for smoothing out the rough edges of market dynamics.

In a rapidly urbanizing global economy and society, a healthy and diverse agricultural sector provides more than just food, feed, fiber, and fuel. The European Union refers to agriculture's multiple benefits as "multifunctionality." Others talk about the "ecosystem services" like ground water storage and carbon sequestration that directly link agriculture with societal and planetary health. While some staunch free market fundamentalists may argue that the marketplace should be the sole arbiter of land use, understanding the importance of and valuing multiple benefits, may be the only way to sustain viable agriculture into the foreseeable future.

A number of benefits are presently undervalued by the conventional costs of food and food production:

Conservation and stewardship. Farmers and ranchers are stewards of more than 50 percent of the lands in the contiguous United States. We simply can't purchase or expropriate all the land needed to protect species and habitats. This is not limited to soils and water resources but includes the stewardship of woodlots, forests, wetlands, grasslands, and complex habitat linkages. Economic assistance and cost-share programs provide incentives to private landowners to maintain habitats.

Rural landscapes. Maintaining land in agriculture can help maintain a rural tax base and provide a stop-gap against development. The aesthetics of a diverse rural landscape, not limited to small farms but including various habitats, enhance property and recreational values of an area. Urbanization of farmland is now underway, paving over, the world's finest topsoils.

Health and homeland security. Maintaining diverse farm sectors will make us stronger and more independent as a nation; we are losing our farm knowledge and the ability to feed ourselves. Bolstering domestic farming capacity reduces dependence on (and vulnerability to) faraway sources of food and fiber.

"Ecosystem services." These include the conservation of large areas of forest

Figure 14

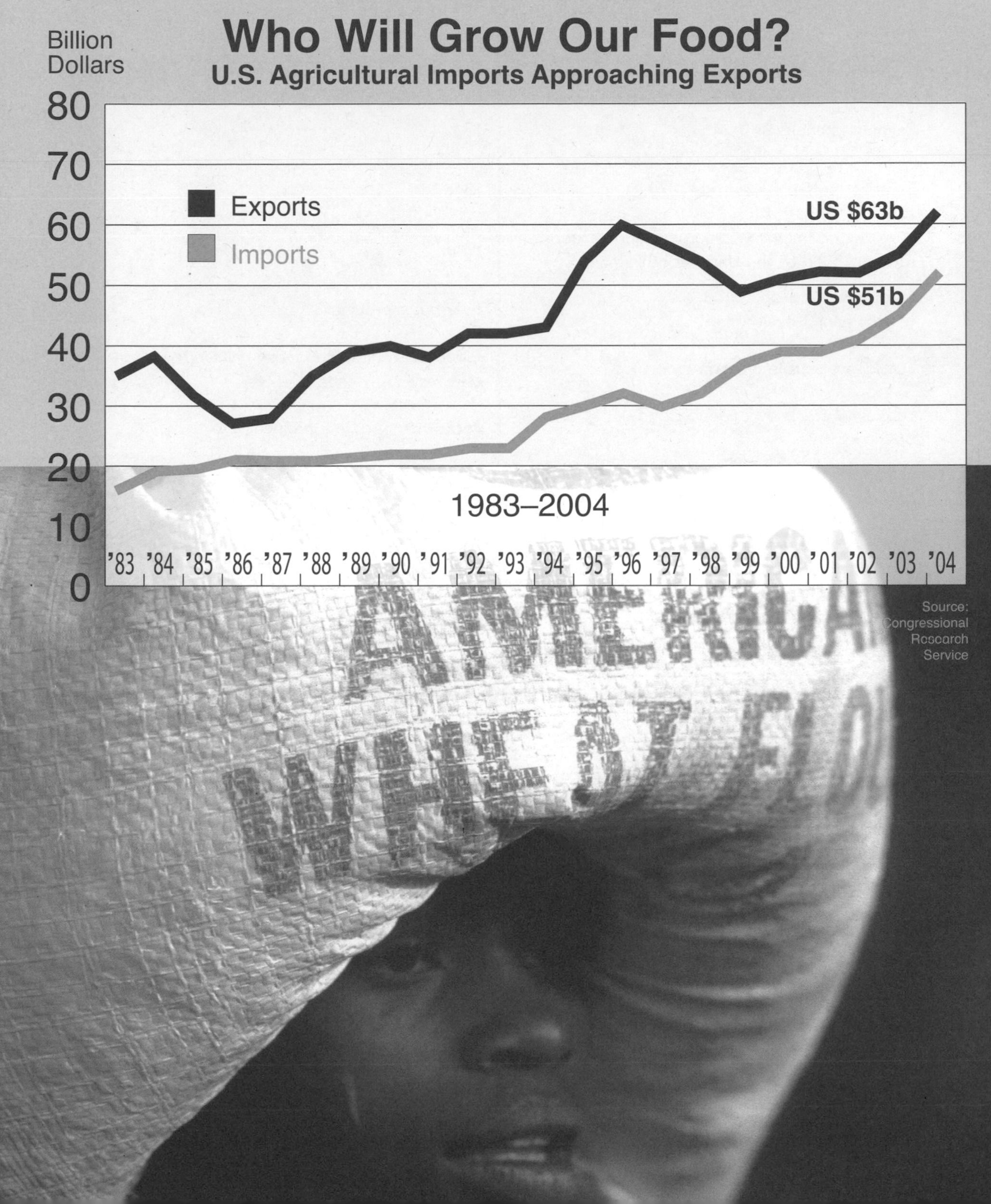
Who Will Grow Our Food?
U.S. Agricultural Imports Approaching Exports
Billion Dollars
80
70
60
50
40
30
20
10
0
Exports
Imports
US $63b
US $51b
1983–2004
'83 '84 '85 '86 '87 '88 '89 '90 '91 '92 '93 '94 '95 '96 '97 '98 '99 '00 '01 '02 '03 '04
Source: Congressional Research Service

Figure 15

Progress Report

Policy Goals	Overall Effectiveness
Payments benefit most farmers	Poor
Subsidies promote fair trade	Poor
Subsidies and nutrition programs support foods recommended by the USDA Food Pyramid	Poor
Programs encourage good environmental stewardship	Poor but improving
Subsidies generate a cheap food supply	Good but with far too many unintended and very costly consequences
Subsidies keep knowledgeable and skillful farmers on the land	Poor—the opposite is happening
Subsides support the external costs of production	Poor
Subsidies achieve fair labor practices	Poor
Food assistance programs and subsidies provide access for all to proper nutrition	Poor
Energy programs support conservation-based efficiency	Underfunded, with too much current emphasis on grain-based ethanol

(which sequester carbon and sustain wildlife), healthy fisheries, buffers against dust and drift, nutrient cycling, water filtration, weather stabilization, attractive aesthetics, natural pest control and pollination, and much more. Our challenge is to find economic mechanisms that provide value for these benefits.[65]

Access to nutritious foods. A healthy food and farm policy can provide much-needed assistance to urban and rural areas, where nearly 40 million Americans who currently go hungry or are forced to spend precious dollars on foods that are high in sugars or fats.

Clean Water Through Healthy Farms

In 1997, the city government of New York realized that because of changing agricultural practices it would need to act to preserve the quality of the city's drinking water. It could have installed new water-filtration plants, but that would have cost $4 billion to $6 billion up front, together with annual running costs of $250 million. Instead, the government pays landowners to preserve the rural nature of the Catskill and Delaware river basins from which New York gets most of its water. It is spending $250 million on buying land to prevent development, and paying farmers $100 million a year to minimize water pollution. Several other American cities, following in New York's footsteps, have calculated that every dollar invested in environmental protection would save anywhere from $7.50 to $200 on the cost of filtration and water-treatment facilities.

Source: *Economist Magazine*, April 22, 2005; For more information, see the New York Watershed Agricultural Council, www.nycwatershed.org.

Should Government Dollars Support Agriculture?

After more than seven decades and more than $600 billion spent on Farm Bill programs since 2000 alone, it seems reasonable to ask: *have these taxpayer dollars accomplished goals of a greater public good?*

A series of inter-related questions may help shed some light on this question:

- What kind of overall food and farming infrastructure do subsidies provide?
- What services does agriculture provide that aren't paid for in the mainstream economy?
- Are the policy objectives still relevant?
- Who benefits most and who loses in the farm policy game?
- Are there unintended consequences of subsidies?
- Do the costs exceed the social benefits?
- Can the objectives be achieved more cost-effectively?

Policy objectives, cautions Mark Ritchie, founder of the Institute for Agriculture and Trade Policy in Minneapolis, may not always be what they seem. When determining the success of a specific outcome of farm policy, Ritchie says it is essential to view the objective through the lens of the ultimate or primary beneficiary. Perhaps an outcome was an accidental failure of a flawed policy. Or perhaps it actually achieved the desired result of some other constituency's very different goal.

Figure 16

Pay Now or Pay Later

Hidden Costs of Industrial Agriculture

WATER CONTAMINATION
Pesticides; Nitrates; Phosphates; Bacteria; Dead Zones

AIR EMISSIONS
Methane; Ammonia; Nitrous Oxide; Carbon Dioxide

BIODIVERSITY LOSS
Wildlife and habitat; Hedgerow and woodlot loss; Bee colony decline; Vanishing Crops and Breeds

Above the Cost of Food at the Checkout Counter

SOIL LOSSES
Erosion; Loss of organic matter and Carbon dioxide

HUMAN HEALTH
Pesticides; Asthma; Bacteria and viral diseases; Antibiotic resistance; Mad Cow and E.Coli; Obesity

DISAPPEARING WETLANDS
Draining and Tiling; Dewatered Rivers; Impact on Species

Source: Based on Jules Pretty, "The Real Costs of Modern Farming," *Resurgence*, Issue 205.

12. World Trade Organization Rulings and the Era of Green Payments

As Americans, we repeatedly hear that trade is a mutually beneficial arrangement for all parties involved, that free trade forms a cornerstone of America's economic policy, and that a world with thriving free enterprise leads to healthy and participatory democracies. It might come as a surprise then that in dealing with agriculture–and commodity-based agriculture in particular–U.S. policies are often tailored to achieve the opposite. While trade negotiators pay lip-service to the ideals of the free market during high-stakes trade talks, inside the Washington Beltway agribusiness lobbyists and Congress use escalating subsidy payments to create an unfair playing field.

Subsidies not only stack the deck toward the biggest operators in our own country, they hurt small farmers in developing countries as well because they boost crop output–thereby lowering global prices. Longtime analyst Peter Rosset from the Institute for Food and Development Policy, describes how subsidies thwart fair trade and food sovereignty in other countries. "Since the 1970s, the U.S. agricultural system has been designed to give American grain-trading giants like Archer Daniels Midland (ADM) and Cargill the edge in capturing the domestic markets of other countries. These companies penetrate Third World markets with a one-two punch. First, the companies work closely with government representatives and negotiators at institutions like the World Trade Organization, World Bank, and International Monetary Fund to force Third World countries to open their markets to American agricultural exports. Once their markets have been forced open, these companies must still be able to out-compete local farmers. To do that, they need an abundant, low-cost supply of commodities, and indeed, they are able to get grain so cheaply that nobody else has any chance of being able to compete, because the second punch is delivered like clockwork every six years by a new Farm Bill designed to depress U.S. farm prices to, and in many cases below, the cost of production."[66]

Brazil successfully argued that U.S. direct payments to cotton farmers were "trade distorting." These payments, the dispute panel found, artificially depress global prices and stimulate overproduction, thereby costing Brazilian cotton farmers millions of dollars in sales.

Nobel Prize–winning economist Joseph Stiglitz takes it a step further. "When subsidies lead to increased production with little increase in consumption, as is typical with agricultural

The Politics of Despair. Voted "Farmer of the Year" in 1989 by the United Nation's Food and Agricultural Organization, Lee Kyung Hae opposed the World Trade Organization's agricultural programs, which he believed destroyed small farms in South Korea, forcing farmers into poverty and despair. On September 10, 2003 Lee Kyung Hae killed himself in protest by knifing himself in the heart during a demonstration at the WTO conference in Cancun, Mexico.

commodities," he explains "higher output translates directly into higher exports. Higher exports translate directly into lower prices for producers. And lower prices translate directly into lower incomes for farmers and more poverty among poor farmers in the Third World."[67] Writing about U.S. cotton export supports in particular, Stiglitz continues, "seldom have so few done so much damage to so many. That damage is all the greater when we consider how America's trade subsidies contributed to the demise of the Development Round."[68]

The United States' (and other industrialized countries') refusal to phase out export subsidies and open domestic agricultural markets has been a primary breaking point in trade negotiations over the past decade. Frustrated by American intransigence, Brazil brought a case before the World Trade Organization arguing that U.S. direct payments to cotton farmers were "trade distorting." On March 3, 2005, a World Trade Organization Appeals Panel upheld a ruling against the United States, concluding that market loans and export subsidies to U.S. cotton farmers artificially depress global prices and stimulate overproduction, thereby costing Brazilian cotton farmers millions of dollars in sales. Similar disputes on U.S. rice and corn supports by other WTO signatories could soon follow.

While commodity subsidies were initially insulated by a "Peace Clause" within the 1992 Uruguay Round of WTO agriculture negotiations, the 2005 Appeals Panel ruling allows Brazil to respond through "cross-sectoral retaliation." In other words, Brazil's attempt to recoup financial losses can take place in business sectors other than cotton or even agriculture–that is, prescription drugs, patent and intellectual property rights, manufacturing, and so on. This is not a small concession. If tariffs or other actions are levied against drugs, movies, software, or automobiles, executives in Hollywood, Silicon Valley, Detroit, and Wall Street could soon take notice of the far-flung consequences of Farm Bill commodity payments.

U.S. commodity subsidies aren't the only violators. Switzerland, Japan, South Korea, and many European governments (France, Norway, and other Group of 10 nations) have been interfering with markets, supporting large corporate agribusinesses, and propping up farm sectors for decades. Theoretically, open markets should help to "float all boats" and create a more prosperous world for all. But trade negotiators of developed countries appear to be deadlocked in a high-stakes game of "chicken." No one wants to be the first country to abandon agriculture subsidies or expose their food markets to the world's lowest cost suppliers by lowering or eliminating tariffs. Yet most agree that the era of artificially supported exports must yield to a more enlightened age of equitable multilateral rules that expand trade but at the same time protect human rights, conserve the environment, and revive family farms and rural communities.

Members of the WTO have already established rules that govern whether agriculture protection and supports are acceptable or not. Referred to as the "Three Boxes" (see Figure 17), these now-expired but still functioning categories establish the legal status of agricultural supports. (1) "Amber Box" subsidies are "trade distorting," and are categorized as supports that illegally fund cheap exports or encourage overproduction. They're not permitted. (2) "Blue Box" subsidies are direct payments to farmers (many of them questionable under current trade rules). These have per-nation ceiling limits. (3) "Green Box" payments are "minimally trade distorting" and the most acceptable under WTO rules. Green Box payments support conservation, rural development, renewable energy, and other investment programs. There are currently no limits on Green Box programs.

The 2005 WTO rulings against U.S. cotton programs (and European sugar beet supports) signal a turning point for agriculture subsidies in industrialized countries. In order to comply with existing WTO rules, the United States needs to trim away billions of dollars in questionable Amber and Blue box payments. According to Peter Riggs, Director of the Forum on Democracy and Trade, violators essentially have three options: (1) do nothing and risk increasing retaliation and further dispute challenges; (2) reduce or eliminate trade distorting subsidies; (3) shift support payments from the Amber and Blue Boxes to rural investment programs acceptable in the Green Box. "The United States is at a crossroads," says Riggs. "We have the choice to live up to our international obligations and invest in ways to make the U.S. more competitive as a whole. Or we can continue giving away money to huge corporations and pay the consequences."

The European Union's Common Agricultural Program (CAP) has already committed to shifting trade distorting export subsidies into the WTO's Green Box. Under the "Decoupling 2013" plan, all tariffs and export refunds on agricultural goods are to be eliminated by 2013. (Whether France and other countries ultimately comply remains to be seen.) Among leading reformers is Great Britain, which launched a national "Environmental Stewardship" program, intended to phase out trade-distorting subsidies and begin paying farmers for the public services they provide, such as wildflower meadows and bird habitats, the restoration of traditional hedgerows, enhanced animal welfare, food safety, and food quality.

In accordance with principles agreed on more than a decade ago, agricultural supports must now move toward regional development, conservation, public health and nutrition, energy security, and other related public benefits. When and how quickly governments realign spending programs or expand market access will no doubt have a fundamental bearing on farmers and rural communities all over the world.

Figure 17

WTO's Three Boxes

Box	Status	Payment Type
Amber	Trade Distorting	• Export credits • Commodity subsidies
Blue	Questionable	• Counter-cyclical payments • Other direct payments
Green	Non-trade Distorting	• Rural development • Green Energy • Conservation

Source: Forum on Democracy and Trade

Trade Talk Timeline

In late 2001 the World Trade Organization began a series of talks—the Doha Round—to assist developing countries by reducing government subsidies and opening markets for agricultural and manufactured goods.

Nov. 2001, Doha, Qatar: The latest round of WTO trade talks starts two months after the 9-11 attacks. Developing countries complain about subsidies, and U.S. cotton payments in particular.

Sept. 2002: Brazil sues the United States through the WTO appeals process, claiming cotton subsidies lower world prices by encouraging overproduction and excess exports.

Sept. 2003, Cancun, Mexico: Talks disintegrate as poor countries protest refusal of affluent countries to reduce or eliminate farm subsidies.

Sept. 2004: WTO rules in Brazil's favor. The United States is required to halt a few direct payment subsidies immediately and to reduce or eliminate certain major subsidy programs.

March 2005: WTO Appeals Panel rules against the United States and sets a timeline for action.

Oct. 2005: Trade negotiators from Washington propose significant subsidy cuts if Europe lowers tariffs and developing countries open markets wider to U.S. exports. No changes materialize.

July 2006, Geneva, Switzerland: Talks are suspended with affluent countries, United States in particular, criticized over its continuation of subsidies.

Aug. 2006: United States discontinues a few subsidies as ordered by WTO. Other challenged subsidy payments remain in place.

Nov. 2006: U.S., European Union, Japan, Australia, India, and Brazil vow to renew talks in 2007.

Source: Dan Chapman, Michael Dabrowa, "Cotton Bailout: An ocean apart, but interwoven, record U.S. exports dampen world cotton prices," *The Atlanta Journal-Constitution*, October 8, 2006.

The Farm Bill Cotton Bail-Out

Will Allen

Every spring American farmers from Georgia to California plant millions of acres of cotton, despite the near certain outcome that by harvest time, their crops will sell at a price far below the costs of production. Although the American textile industry has long been in decline, U.S. cotton farmers plant vast acreages anyway, assured that billions of dollars in taxpayer subsidies will help get their fiber to export markets. At least that's the way it has worked under recent Farm Bill arrangements.[69]

In 1990, the United States had approximately 40,000 cotton farmers, primarily in the Deep South, Texas, California, Arizona, and New Mexico. They harvested almost 12 million acres of cotton and produced 15.5 million 480-pound, refrigerator-sized bales. The price for upland cotton averaged 68.2 cents per pound of lint.[70]

Just five years later, their numbers had fallen to nearly 30,000, while both land area and output climbed to 16 million acres and 18 million bales. The average price for U.S. cotton rose to 76.5 cents per pound of lint. By 2001, the number of cotton farmers declined yet another 25 percent, to 22,000. Planted acreage dropped to 14 million acres even while output spiked to more than 20 million bales. The average price for upland cotton plummeted to 35.1 cents per pound.[71]

While the number of U.S. cotton farmers dropped by more than one-half between 1990 and 2005, the amount of farmland essentially remained the same—averaging around 12 million total acres per year. As the price dropped for cotton fiber due to an excess world supply between 1995 and 2002, nearly all but the most capitalized and subsidized U.S. growers stopped planting cotton.

Nearly three-fourths of U.S. cotton farms are now leased by absentee and live-in landowners, so it is probably not surprising that land has become increasingly concentrated in the hands of fewer large-scale farmer-renters.

When prices are low, landlords want to be sure that the farmer who manages their land has the best chance of delivering high yields. This is because historical yields determine the amount of the subsidy payments, which the landlord normally shares with the farm manager. As a result, landlords tend to go with the managers who have the newest equipment and the most cash reserves to hire labor. Small- and medium-scale farmers are frozen

world trade by the World Bank, the International Monetary Fund, Oxfam, and several other development and free trade advocates. Cotton farmers from India, West Africa, Brazil, and many other countries couldn't sell their cotton abroad for a profit, and they couldn't compete at home against cheap U.S. cotton. More than two dozen countries dependent upon cotton sales for important foreign exchange earnings

out of the bidding for leased land.

As smaller cotton farms "failed," many large-scale farmers gobbled up their acreage. While prices for cotton plummeted between 1995 and 2002, U.S. cotton subsidies increased more than ten-fold—1,020 percent—thanks to the Commodity titles of the 1996 and 2002 Farm Bills. Nearly 20,000 farmers still grow cotton in the United States, yet only a small club of growers get most of the subsidies. In 2001, for example, when annual cotton subsidy payments reached at least $3 billion, less than 10 percent of the farmers received more than 80 percent of those subsidy payments. This meant that fewer than 2,000 cotton farmers got somewhere around $2.4 billion.[72] That's an average payment of $1.2 million per "farmer."

When the price of cotton skidded from 76.5 cents to 35 cents and finally hit bottom at 28 cents it was not only the small- and medium-sized U.S. farms that fell by the wayside. As world cotton prices fell, many large cotton mills in the United States also closed their doors. Meanwhile cotton brokers dumped their surplus and cheap cotton on foreign countries that had been lured into

suffered severely.

In the last ten years, an estimated 40,000 cotton farmers in India have committed suicide. Surviving family members testified at a Bangalore tribunal in 2000 that their husbands, sons, and brothers committed suicide because they couldn't pay their seed or pesticide or fertilizer bills and were going to lose their land. While many debt-ridden farmers on the Indian subcontinent tragically chose to end their own lives, thousands more sold their kidneys on the international organ market instead of committing suicide. These farmers received about US$800 for their kidneys, which usually went

U.S. Cotton Subsidies and U.S. Cotton Exports, 1996 and 2005

	1996	2005
U.S. subsidies (billions)	$0.6	$3.4
U.S. exports (millions of bales)	6.9	17.6

Sources: USDA; Oxfam America.

Top exporters of cotton 2005-2006

(Thousands of tons)

Country	Thousands of tons
United States	4,212
Uzbekistan	1,128
West Africa	899
India	780
Australia	720
Brazil	473

to pay their bills and stave off bankruptcy.

The West African cotton industry hasn't fared much better. Oxfam estimates that more than 10 million West Africans were economically devastated by the drop in cotton prices and the dumping of subsidized U.S. cotton. A common African lament at the World Social Forum in Mumbai, India in 2004 was: "The more we produce, The more we export, The poorer we get."

After complaining for several years about the damage from U.S. cotton subsidies, Brazil finally filed a suit in the World Trade Organization (WTO). In 2005, Brazil won a judgment after the United States lost two appeals of the decision. The WTO judicial body determined that the U.S. cotton subsidy program was in violation of world standards for free trade.

Theoretically, the United States is supposed to change its practice of subsidizing surplus cotton since

Brazil won its judgment. But history suggests otherwise. Ever since the Civil War, the U.S. cotton elite has usually figured some way around most governmental and international roadblocks. The Doha Round of WTO negotiations dissolved in 2006 over disputes about agricultural subsidies. Cotton was one of the critical pieces of the negotiations and disputes. King Cotton, in the form of Cotton Incorporated and the Cotton Council, worked publicly to prevent the signing of any free-trade agreement that limited subsidy payments for cotton farmers. While they claimed an interest in protecting all cotton farmers, their primary concern was a small number of very rich farmers who receive 80 percent of the taxpayer subsidies.

Will Allen received his Ph.D. in Anthropology from the University of California Santa Barbara and is a life-long farmer. He is the former director of the Sustainable Cotton Project.

13. New Zealand: Still Subsidy-free After All These Years

In 1984, the government of New Zealand announced the unthinkable. Faced with mounting fiscal deficits and spiraling inflation, the country abandoned its decades-old extensive programs that had been cushioning farmers with as much as 40 percent of total income through the 1970s and early 80s.

At the time, New Zealand's farmers faced a crisis that was not too radically different from the shocks affecting U.S. agriculture during the 1970s and 80s. Rising oil prices had triggered inflation, and farmers were finding it increasingly difficult to recoup fair prices on the open market. Great Britain, historically the country's most secure trading partner, had joined the European Union, and was in the process of realigning its trading relationships. Once privy to special status as a Commonwealth nation, the island country's agricultural exports were cast into the already overflowing basket of the global economy.

New Zealand's farm leaders brought national attention to the need for urgent subsidy reform.

The New Zealand government initially responded to these challenges the way most developed countries still do today: by shoveling money and financial incentives at the farm sector in the hope that boosting production would also boost farm income. Tax breaks, fertilizer subsidies, price supports, low-interest loans, disaster relief, weed-eradication payments, special training programs, and other farmer-friendly incentives unfortunately all funneled toward a similar result. Rather than fostering greater returns, they generated oversupply, which in turn, depressed commodity prices.

New Zealand farmers were quick to catch on to the flawed logic of the system, even as it worked to their advantage. They dubbed the "livestock incentive scheme" a direct payment program to help farmers increase the size of their herds–the "skinny sheep scheme."[73] (Even a pasture-rich country falls victim to such fundamental laws as stock-carrying capacity.) One farmer comically named his newly purchased boat "the SMP," after the "supplementary minimum prices" subsidy, the backbone of the country's farm supports.[74]

Ironically, however, New Zealand's farm leaders brought national attention to the need for urgent subsidy reform. It had become obvious that this growing catalog of farm support programs was overburdening the national treasury, and the subsidies weren't achieving what

A Subsidy-free View. More than 20 years after New Zealand's Labor Party shut down an extensive agricultural support program, the farm sector is overachieving. Sheep stocking numbers have declined but have been replaced with a diversity of activities. Agricultural productivity, while averaging 1 percent growth per year in the subsidy era, has reportedly grown 5.9 percent per year since 1986 when farm supports were abandoned.

they were supposed to: compensating farmers for the inflationary pressures that were causing production costs to escalate. By 1982, the country's leading farm organization, the Federated Farmers of New Zealand, advocated a partial overhaul of the subsidy system. Then in 1984 the country's Labor Party won a landslide victory. Included in a sweeping package of economic restructuring designed to curb inflation and minimize debt was the elimination of nearly all agriculture programs.

While certain government safety net programs remained in place, New Zealand's farm sector was forced to go it alone. By most accounts, the initial six-year transition was a rocky one and especially difficult for farmers with unpaid for equipment and improvements. An estimated 1 percent, or 800 farmers, were forced to leave the land. But this was still far fewer than the 8,000 farmers or 10 percent that had originally been predicted to go. Sheep farmers, the most heavily subsidized before the reforms,

Why New Zealanders Don't Like Subsidies

- Resentment among farmers, some of whom will inevitably feel that subsidies are applied unfairly.
- Resentment among nonfarmers, who pay for the system once in the form of taxes and a second time in the form of higher food prices.
- The encouragement of overproduction, which then drives down prices and requires more subsidization of farmers' incomes.
- The related encouragement to farm marginal lands, with resulting environmental degradation.
- The fact that most subsidy money passes quickly from farmers to farm suppliers, processors, and other related sectors, again negating the intended effect of supporting farmers.
- Additional market distortions, such as the inflation of land values based on production incentives or cheap loans.
- Various bureaucratic insanities, such as paying farmers to install conservation measures like hedgerows and wetlands—after having paid them to rip them out a generation ago, while those farmers who have maintained such landscape and wildlife features all along get nothing.

Removing subsidies, on the other hand, forces farmers and farm-related industries to become more efficient, to diversify, to follow and anticipate the market. It gives farmers more independence, and gains them more respect. It leaves more government money to pay for other types of social services, like education and health care.

Excerpted with permission from Laura Sayre, "Farming without subsidies? Some lessons from New Zealand," *The New Farm*, March 2003.

bore the brunt of the impact, as did operations that were heavily in debt. Within half a decade, however, a recovery was well underway. By the early 1990s, land values, commodity prices, farm profitability, and other indices had stabilized or even begun to show steady improvement.

According to most reports, rather than a collapse of agriculture on the island, New Zealand's shift has led to an energizing transformation of the food and farming sectors. Profitability, innovation, and agricultural diversity have returned to farming even without extensive direct production subsidies. While approximately 90 percent of New Zealand's total farm output is exported, most of the food consumed inside the country is grown domestically. The total number of sheep have fallen, but weight-gain and lambing productivity have increased. Similar efficiencies have been experienced in other farm sectors. By 2000, New Zealand's dairy industry was earning more foreign exchange than sheep farming and its production costs were among the lowest in the world. Barely in its infancy in the pre-1984 era, New Zealand has developed a thriving wine industry.

What's more, a whole new generation of New Zealanders has entered the farming industry–a generation that has little to no knowledge about subsidies. In a 2001 BBC News interview, Alistair Poluson, chairman of the Federated Farmers of New Zealand offered this advice to developed world farmers based on the country's experience: "Get off the subsidy gravy train as soon as possible."[75]

One might argue that New Zealand represents a unique case in the world of farm subsidies, because of its abundant resources, relatively small population, and geographic distribution. Another sound argument can be made for the country's social benefits such as public health care, education, and other services that subsidies indirectly helpt to compensate U.S. farmers for. But with the country now entering its third full decade without subsidies, one thing that can't be said is that the New Zealand subsidy reduction policy is "experimental" or part of an ongoing research trial.

While countries around the Pacific and around the world are emptying government coffers with output incentives, subsidized water and electricity rates, and countless other programs for their farm sectors, current government financial support for farmers is less than 1 percent of average farm income across New Zealand.[76] Farm output and farm income are on the rise. Agriculture contributes slightly more to the economy than it did during the subsidy era and has taken on a culture of creativity and entrepreneurship.[77]

The chief architect of New Zealand's free market initiative, Roger Douglas–initially vilified–has been knighted.

What would agriculture look like with far fewer subsidies and direct payments than we have now? There is actually a 20-year case study with a not too dismal ending. Apparently many countries are taking careful notice.

WEDGE

ISSUES

14. Wedge Issue Politics

It would be naïve to imagine that the Farm Bill will be overhauled during any single negotiation cycle (New Zealand style) to spontaneously create the food and policy systems we need or deserve to have. The balance of power between the agribusiness lobby, the relative overrepresentation of farm states with declining rural populations, public interest groups, and the inertia of entitlement programs is enough to turn most optimists into dyed-in-the-wool skeptics. One long-time resource economist describes the Farm Bill as a "fully-rigged" game. Somehow hope must eternally spring forth.

Considering everything at stake–our health, our food, our environment, our children's future–one might think that the forces opposing big agriculture could build bridges and form an unstoppable united front for change. But more often than not, opponents of commodity subsidies have been divided by their own narrow issues. Too frequently, the concessions that reformers have won through hard-fought campaigns–from conservationists, to family farm advocates, to antihunger groups–only serve to make an unsatisfactory system slightly "less bad."

Politics is a reactive arena, however, and over seven decades of Farm Bill negotiations, some watershed shifts in program commitments have occurred as a result of successful campaigns. Once established, these interests–wedge issues, if you will–become assumed by the ever-expanding Farm Bill pie. Wedges become titles and later entitlements, over time demanding more money, sometimes raiding the till from other programs. Eventually taking back these entitlements from constituencies becomes difficult but not impossible, even after relevance, appropriateness, or effectiveness have passed. (Peanut and tobacco farmers have been recently bought out, for example.) Or sometimes agribusiness merely finds ways to co-opt programs to their advantage, speaking the language of public benefit, while funneling the money to Big Ag and mega-farms. (Channeling conservation dollars to feedlots for sewage containment or providing subsidies to venture capital–rich corn ethanol plants both come to mind.)

At least two major forces have redirected the course of the Farm Bill since its origins as a set of extreme measures to address the needs of millions of farmers and millions of hungry and unemployed citizens. In 1961, the food stamp program was reinstated in the early days of the Kennedy administration. Here, commodity agriculture met an emerging

Congressional hunger caucus, tying farm surpluses and agricultural output to a malnutrition crisis and the need for a welfare safety net for the poor. Then in the mid-1980s, conservation resurfaced as an important Farm Bill priority. Environmental and wildlife advocates successfully lobbied for stewardship incentives that went beyond erosion control–the primary and nearly sole focus of early Dust Bowl conservation programs.

As an old saying goes, change is the only constant. Numerous circumstances are aligning in the first decade of the twenty-first century that make food and farm policy reform an absolute necessity.

- Mounting health care costs due to adult and childhood obesity;
- Rising costs and eventual limited availability of fossil fuels;
- Decline of the U.S. agricultural trade surplus;
- Brazilian victory in the World Trade Organization cotton case;
- Major push for farm program payment limitation reform;
- Extreme weather events due to global warming;
- BSE (Mad-cow disease), *E. coli* contamination, and other issues threatening the long-term viability of industrial feedlots.

Energy. With rapidly fluctuating oil markets and unpredictable climate patterns upon us, energy issues will dominate Farm Bill debates well into the future. The corn-based ethanol industry has enjoyed steady growth thanks to generous tax incentives and other subsidies over the past twenty years. Now a major push is underway to replace a significant amount of imported petroleum with biofuels derived from fields, grasslands, tree farms, and even forests. Whether the ethanol wedge causes the Energy title and the already bloated Commodity title to expand or cannibalize the consistently short-changed Conservation title, (or both) is a rapidly developing topic. These complex and inter-related issues will be discussed at length.

Health Care. Perhaps the most influential lobbying force poised to weigh in on Farm Bill discussions is the health care community. The direct link between subsidized high-calorie processed foods and the epidemic of diet-related maladies has reached the desks of governors, state and federal health administrators, doctors and nurses, parents and teachers, health insurers, and chief executive officers throughout the country. With the annual medical costs of the obesity crisis now exceeding the yearly $90 billion Farm Bill, the health care community literally represents a sleeping giant with the ability to throw its significant influence and resources behind food and farm policies. Will the health care community demand changes in agricultural policy to coincide with nutritional and dietary changes that are so obviously necessary? Will the health care community join forces with other movements calling for food production and distribution arrangements that center around regions feeding regions and leadership in organic production methods and grassfed livestock farming?

Religious Communities. The faith-based community is yet another constituency whose influence remains fairly untapped. According to Thomas Forster of the Community Food Security Coalition in Washington D.C., organizations such as the National Catholic Rural Life Conference, Bread for the World, and many others that have long worked on issues of fair trade, antihunger, and the devastating impacts of agribusiness consolidation on small farmers and rural communities now well understand the far-reaching implications of food and farm policy. Conditions of animal confinement, the untoward power of monopolies, and the ethics of dumping cheap agricultural products on the world's poor have surfaced as moral concerns among many spiritual communities. Broader participation among church members and faith-based organizations could greatly expand citizen input into food and farm politics, particularly outside traditional agricultural regions.

Wedge Issue Politics? To what extent these and other issues and formerly silent constituencies will affect the direction, vision, and general balance of the Farm Bill pie remains to be seen. Policy changes come slowly and idiosyncratically, often taking years and multiple Farm Bill cycles for even a single issue or new program to take root. Changes rarely address the systemic and interconnected nature of key problems. Will reform arrive slowly in the form of a massive shift to green payments? Will some critical emergency shock the public into action, indignation, or outrage? Who knows. One can only hope today's reform movements recognize the urgency and importance of bridge building and both the favorable and unfavorable consequences of wedge issue politics.

15. Hunger, Health, and Nutrition: Changing the Policy Palette

Food—its cultivation, preparation, and enjoyment—remains our most fundamental connection with the earth. Clearly early Farm Bill policies attempted to bridge that connection by linking the country's agricultural output with the nutrition demands of those in need. Seven decades later, despite benefiting from the world's most technologically intensive food and farming system, and spending less income on food relative to many other nations, a majority of Americans do not enjoy healthy diets. Eating a diet high in calories doesn't necessarily ensure that one is well fed—even if that food is "cheap." Nor does subsidizing vast quantities of commodity feed grains, oil seeds, and antibiotic- and hormone-laced meat and dairy products guarantee that tens of millions of adults and children will be spared hunger or food insecurity.

Here are just a few troubling signs of imbalance with the country's farm and food policy:

- Only 2 percent of 2- to 19-year-olds meet all five federal requirements for a healthy diet.
- In 2000, five "vegetables"—lettuce, frozen potatoes, fresh potatoes, potato chips, and canned tomatoes—made up almost half of the total vegetable servings in the United States.
- The food industry spends $15 billion a year marketing to children. The Federal School Lunch Program spends only $7 billion per year to feed our children in the public schools.
- The average American consumes more than 50 gallons of carbonated soft drinks every year.[78]
- Nearly 12 percent of Americans are "food insecure," or experience "relatively low food security," USDA shorthand meaning that they are often not certain where their next meal will come from.[79]
- Due to a rise in obesity and type-2 diabetes, this generation may be the first in American history to die at a younger age than their parents.

Figure 18

Calories In, Calories Out

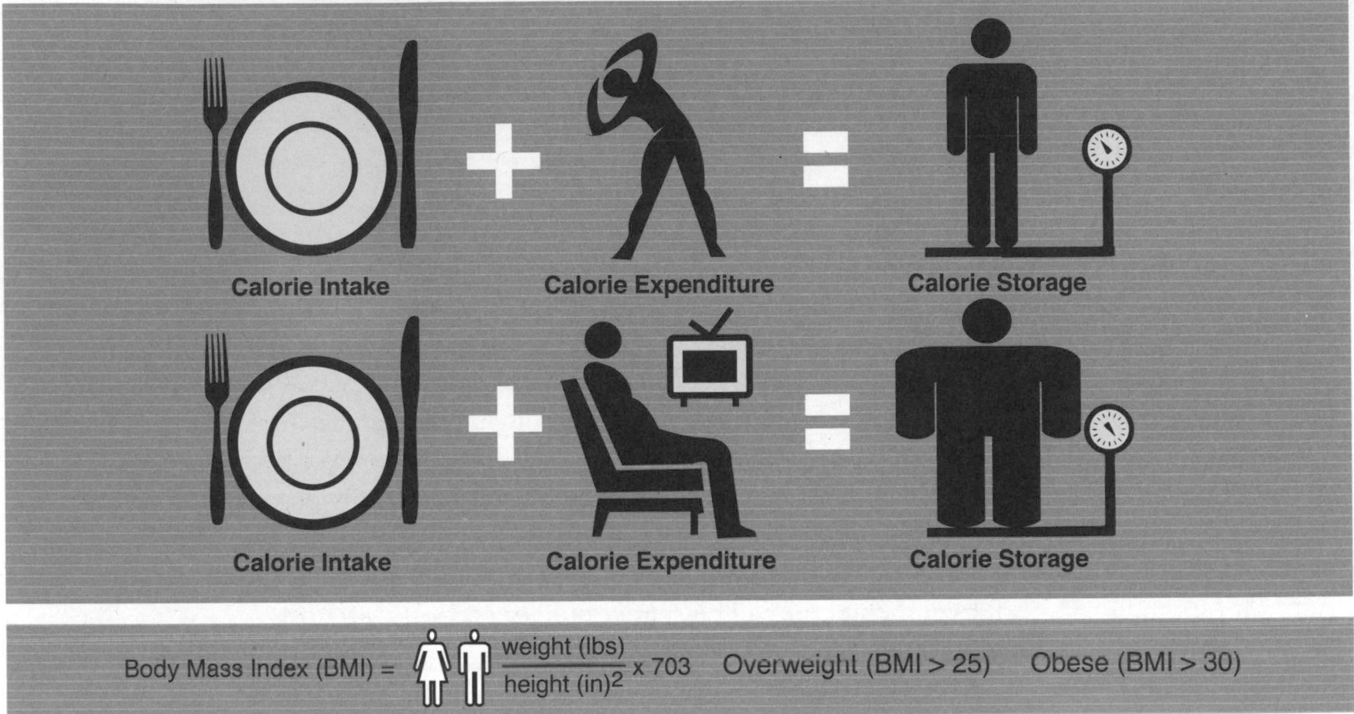

Obesity: America's Heavy Burden

In 2001, the Surgeon General sounded the alarm, declaring that U.S. obesity rates–double those of the early 1970s–had reached "epidemic" dimensions.[80] A year later, researchers tallied that two out of every three adult Americans were clinically overweight: 31 percent of adults between the ages of 20 and 74 were medically obese (more than 100 pounds overweight), raising their susceptibility to heart disease, diabetes, and other ailments. Another 34 percent were at least 30 pounds over the recommended body mass index.[81] Perhaps most troubling, the percentage of overweight children tripled during the same period. (See Figure 21, *Overweight and Obesity on the Rise.*)

A smorgasbord of influences is driving the country's excessive weight gain. Insufficient exercise and sedentary lifestyles. Increasing dependence on meals eaten on the run or outside the home. Unhealthy foods in school cafeterias. Predatory marketing and super-sizing of junk foods and processed foods.[82] A genetic predisposition known as the "thrifty gene" that causes humans to gorge on available foods regardless of the nutritional consequences.[83] While this crisis transcends race, class, and gender, it is clear that poor ethnic communities are being disproportionately impacted.

What the Surgeon General didn't reveal, however, is the leading role that U.S. food and farm policy plays in our nation's expanding waistlines. Federal policies send dangerous ripple effects throughout the food chain. It starts with corn and soybeans, storable crops that serve double duty as both cattle feed and a source of processed sugars and hydrogenated fats. As Farm Bill subsidies have lowered prices of commodity crops over the past thirty years, the food industry has invested heavily in an infrastructure that turns cheap materials into highly profitable "value-added" products.

High-fructose corn syrup (HFCS), not even commercially concocted before the 1970s, has rapidly become one of the Farm Bill's industrial workhorses. A liquid sweetener with six times the potency and far cheaper than cane sugar, HFCS can also be used to prolong shelf-life, resist freezer burn, create an oven-toasted effect, and other processing functions.[84] Over the past three decades, U.S. consumption of high-fructose corn syrup has jumped 1,000 percent. Soybeans are almost as versitle, providing a cheap and abundant source of added fats in the form of hydrogenated oils that have almost invisibly worked their way into the makeup of nearly every nonproduce item in the modern industrial diet. Dairy and meat products, made from livestock raised in confinement conditions and fed rapid weight-gaining diets of corn and soybeans are also high in unhealthy fats. The utility and commercial desirability of these ingredients is obvious. Out of the nearly 15,000 new food products introduced each year, 75 percent are candies, condiments, breakfast cereals, baked goods, beverages, or dairy novelties.[85]

Along the way the average U.S. citizen's daily food intake has ballooned to nearly 3,900 calories–almost twice the maximum recommended by U.S. health officials.[86] This includes, on average, 32 teaspoons of added caloric sweeteners per day and as high as 1,800 calories in fats.

What About the Food Pyramid?

It's not that difficult to draw a correlation between Farm Bill subsidies and the country's dietary imbalances. The federal nutritional guidelines in the USDA's food pyramid actually recommend reducing added sugars and fats–the same ingredients made cheaper by government subsidies. While the costs of corn- and soy-based processed food ingredients have plummeted and their proliferation in the food system has spread widely, the prices of regionally grown healthier foods recommended by the USDA–fresh fruits and vegetables, in particular–have risen sharply.

A recent study by Heather Schoonover and Mark Muller at the Institute for Agriculture and Trade Policy in Minneapolis, shows that the costs of fresh fruits and vegetables, grown with little federal support, increased nearly 40 percent between 1985 and 2000. Meanwhile, the price of soda and soft drinks decreased by almost 25 percent over the same time period. (See Figure 20, *The Higher Cost of Healthy Foods.*)

And when it comes to choosing healthy foods, price matters. Researchers at the University of Minnesota, for example, found that reducing prices even in small increments can significantly influence purchases of fresh fruits and vegetables. And decreasing the costs of fruit and salad by 50 percent resulted in a four-fold increase in sales for one cafeteria studied.[87]

Food Deserts: So Much Agriculture, So Little Food

The Farm Bill's structural distortions reach far beyond a serious dietary crisis. Based on a cheap materials policy, three decades of investments in farming, processing, retailing, and distribution have created nutritional deserts throughout our regional economies.

The lack of government support for produce crops makes growing healthier foods a far riskier proposition for farmers. Even if they want to diversify, farmers often face rural economic infrastructures largely tailored toward commodity production and lacking in necessary equipment, supplies, technical expertise, cold storage warehouses, and slaughter and processing facilities. Regional food distribution chains that once accommodated a diversity of crops have been entirely eliminated in many traditional farming areas.

This partly explains the paradox of "so much agriculture, so little food." Commodity farms lose money by the millions growing feed grains, cotton, and oil seed crops, only to be compensated for those losses by federal subsidies. Meanwhile, local food stores offer fruits and vegetables supplied year-round by low-cost producers sourced outside the region–often from across the country and from around the world. In fact, fruits and vegetables are the fastest growing category of U.S. food imports, increasing at an average annual growth rate of 8.4 percent.[88] Most of the cash residents spend on their weekly food bills leave the region and the state ("value-leaving"). Dinner tables remain disconnected from the fields that surround them, giving rise to "food deserts."

Food deserts aren't strictly a rural phenomenon either. Many inner-city urban areas, particularly low-income neighborhoods, have become "underserved markets," where it is often easier to find a fast food restaurant or a convenience store than a grocery store with a variety of more healthy options. According to Adam Drenowski, professor of epidemiology at the University of Washington, people are gaining weight and getting sick because unhealthy food is cheaper and often more available than healthy food.[89]

If expanding access and availability of healthy foods throughout the country becomes a Farm Bill goal, as it should, spending priorities must be drastically altered to bring about structural changes. Consider for example, that over the course of the 2002 Farm Bill, $15 to $23 billion has been spent each year on commodity crops. In 2005, less than $1 million–not even a thousandth of a percent of that sum–was spent to promote the country's 3,700 farmers' markets that

Figure 19

What About the Food Pyramid?

Daily Reality

HIGH-FRUCTOSE CORN SYRUP | HYDROGENATED FATS | CONFINEMENT DAIRY & MEAT

Daily Recommended

GRAINS | VEGETABLES | FRUITS | MILK | MEAT AND BEANS

The Higher Cost of Healthy Foods

Change in Food Prices Over a 15-year Period, 1985–2000 (Real Dollars)

Source: *USDA ERS Food Review*, Vol. 25, Issue 3. Converted to real dollars and published in "Food Without Thought: How U.S. Farm Bill Policy Contributes to Obesity," by the Institute for Agriculture and Trade Policy, 2006.

Figure 21

Overweight and Obesity on the Rise

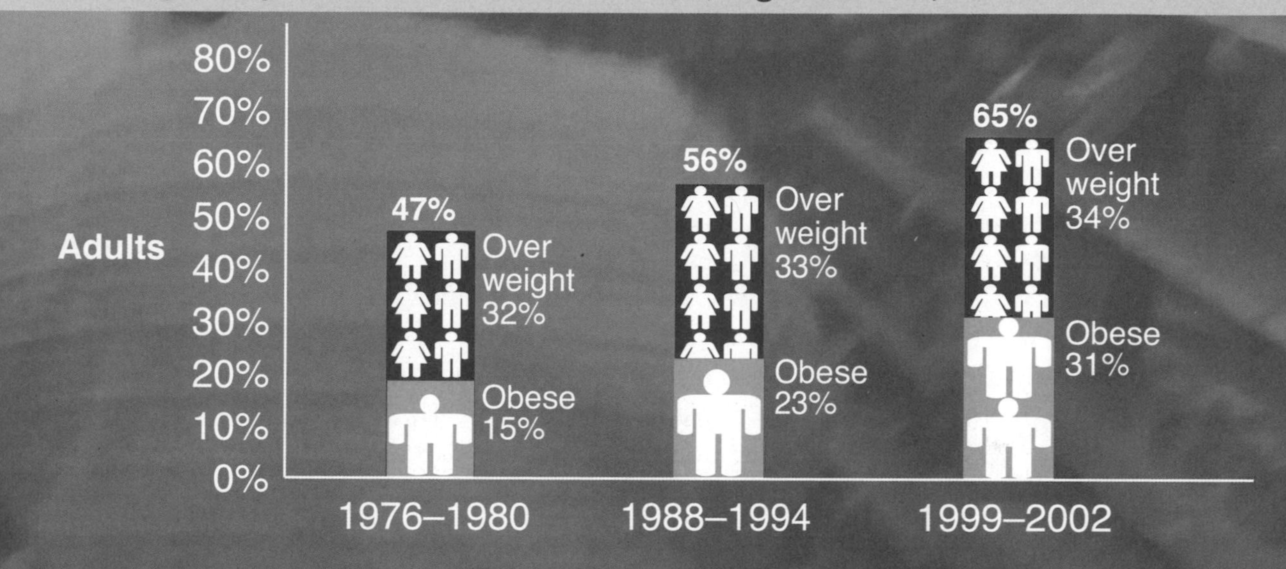

Sources Adults: National Center for Health Statistics; Census Bureau; Nielson & Popkin, 2004, Associated Press; Children & Adolescents: Center for Disease Control/National Center for Health Statistics, National Health Examination Survey and National Health and Nutrition Examination Survey

Body Mass Index (BMI) = $\frac{\text{weight (lbs)}}{\text{height (in)}^2} \times 703$ Overweight (BMI > 25) Obese (BMI > 30)

provide access to locally grown produce to millions of citizens.[90]

Calories versus Nutrients: The Antihunger Community's Catch 22

For more than 30 million Americans, the Farm Bill Nutrition title often means the difference between going to bed hungry and receiving minimal nourishment. Food stamps, Women, Infants, and Children's (WIC) vouchers, free and reduced-price school breakfasts and lunches all help to enhance the well-being of low-income Americans. To their great credit, Farm Bill after Farm Bill, the antihunger lobby has heroically defended the country's hunger safety net, attaining the necessary votes to "do no harm" to food assistance programs that counter a policy of low-wages. The antihunger lobby has also helped to destigmatize the experience of food assistance. An innovative credit card system (Electronic Benefits Transfer, EBT) has replaced paper coupons and can even be accepted in some farmers' markets.

Yet it is one of the Farm Bill's profound ironies that the costs of maintaining food stamps and other assistance programs have also contributed to the degradation of our overall food system. Political survival for the antihunger camp has pivoted around an ongoing alliance with the commodity lobby. While researchers have not conclusively demonstrated that food stamps contribute directly to obesity,[91] the Farm Bill's hunger safety net, by its very structural ties to commodity-based processed foods, has become a calorie delivery system rather than a nutrition program. The country's poorest communities and populations have little choice but to spend their dollars on the cheapest foods available. And on a per-calorie basis, energy-dense foods–those high sugars and fats–are cheapest and tend to be the least nutritious.[92]

It is also a painful irony that obesity can affect the poor, the hungry, and the food insecure even harder than affluent and educated populations who can afford to buy more food. The expense of transportation to a good market, the pressures of monthly cash flows, the stresses of coping with uncertainty and long-term poverty can all impact dietary choices and negatively impact eating behavior. Tragically, many nutritional problems that affect the young can be lifelong. "It doesn't take much nutritional depravation to retard a child's developmental growth," laments Mark Winne, long-time researcher for the Community Food Security Coalition.

The Health Care Costs of "Cheap Food"

The U.S. Surgeon General estimates that Americans now spend more than $100 billion annually on illnesses caused by obesity–even more than annual Farm Bill budgets. An increasing percentage of these costs are footed by taxpayers. And they are affecting our ability to lead productive lives. In 2006 alone, obesity-related illness were responsible for 40 million lost work days, 63 million doctors visits, 239 million restricted activity days, and 90 million bed-bound days.[93]

These nutritional diseases should be preventable. At least one miraculously simple counter-reaction could help stuff at least some of these high-calorie sugar and fat genies back into their bottles. Farm and food subsidy programs could be realigned to support the federal dietary guidelines and oriented toward food chains that produce and distribute locally grown, healthy foods. But it will take enormous input from the health care community and other large institutional players to transform high-calorie industrial food deserts into diverse nutrition-rich regional food chains.

With refocused priorities, however, billions of dollars in annual Farm Bill payments could jump-start such a transformation of the food system. Institutional markets—from school and university cafeterias to hospitals and government agencies and businesses of all walks—can bring enormous additional food purchasing power to bear. Of course, this will mean shifting from "low-bid" procurement policies toward values other than the lowest price. With larger budgets and an expanded mission, procurement departments could set a whole new food chain in motion by favoring fresh items, regional producers, reduced Food Miles, and sustainable farming partners.

Sea changes in the public behavior around health are not unprecedented. In the mid-1960s, the Surgeon General's warning against cigarette smoking changed attitudes about the long-term health effects of tobacco use. The President's Council on Physical Fitness established new protocols and standards for public schools in the 1970s. There is no reason a healthy food and fitness movement could not radiate through all levels of society. Farmers markets—double or triple the nearly 4,000 in existence today—could serve as hubs of healthy food education and distribution.

Replacing our current high-calorie bill with a Food and Farm Bill that promotes healthy lands and healthy people requires the willingness to stand up and demand that we want to feed our children and ourselves a diet that meets the nutritional and environmental challenges of the twenty-first century. Hunger, health, and nutrition just might deliver the political wedge that redirects us toward that fundamental connection with the earth.

Farm-Direct Nutrition

Delivering Fresh Produce to Seniors in Need

Deborah Rich

When the City of Oakland's Commission on Aging solicited proposals in 2003 for programs to improve the nutrition of low-income seniors in the community, Suzan Bateson, executive director of the Alameda County Community Food Bank, responded with an idea simple in conception yet complex in execution: provide the seniors with dependable access to more and higher quality fresh produce at low cost.

With a little over $32,000 in grant money from the Commission on Aging in hand, the food bank contracted with Full Belly Farm, a 200-acre certified organic farm located one hour northwest of Sacramento, to make monthly deliveries of produce directly to three of the food bank's member organizations which were already providing food assistance to low-income seniors in Oakland. The grant money also paid for deliveries to Mercy Brown Bag, a program sponsored by the Sisters of Mercy that makes bags of groceries available to low-income seniors.

The direct link forged by the food bank between the farm and the food assistance programs helped to keep costs down and the quality of the produce high by reducing handling and transit time. For the past two years, the program has provided 1,500 low-income Oakland seniors, 200 of whom are homebound, with at least three produce items per month at an average cost of $1.65 per senior per delivery. Full Belly harvests on one day and trucks the carrots, chard, and lettuce to each distribution center in Oakland the next. The seniors generally have the produce in their kitchens by the following day.

"What the Commission on Aging was looking for, and what we were looking for, was a way to deliver good nutrition to seniors in need. We chose produce because produce is expensive and hard to get; a lot of times individuals don't have the resources to go to grocery stores and select what they want," Bateson said.

But, the Alameda County Food Bank does not have funding to continue the contract with Full Belly Farm during the 2005–2006 fiscal year. A last-minute complication in the grant solicitation process caused the food bank to miss the city's deadline.

"This year the Commission on Aging decided as part of their (request for proposal) process to have organizations develop collaborative partners. We thought we had identified a collaborative partner, and were ready to submit. But at the end of the day, the partner did not choose to sign off on this project," Bateson said. "We are actively looking for another funder that has interest in this project."

The food bank has always had the goal of offering produce to individuals in need, but it generally had to take what it could get, where it could get it, without having the luxury of specifying the quantities and freshness. "These deliveries (secured by the grant funds) are a guarantee," Bateson said. "When you're dealing with donated commodities, you don't have a guarantee."

Bateson knew that if she could get the produce delivered, by and large the seniors would welcome it, know how to prepare it, and enjoy eating it.

"With seniors, it seems that any fruits and vegetables that we come up with, they remember preparing them from their childhood on," Bateson said.

For Alma Ferguson, 73, washing and cooking fruits and vegetables has been a fact of life.

"If you want to eat, you have to do the preparation," she said. A volunteer with Mercy Brown Bag, Ferguson helps assemble and distribute bags of groceries for fellow low-income seniors.

"I haven't heard of one person that isn't glad about the produce. And I'm there when the bags are passed out."

Norman Rollins, 75, lives alone. His grandparents raised him on a 300-acre corn, wheat, and hog farm in Illinois, where his grandmother, a Mohawk Indian, grew all the family's vegetables in a garden about half the size of a town lot. She canned "everything she needed," and reserved her store purchases for sugar, flour, and spices.

"This produce helps me out quite a bit," Rollins said. "I'm on a fixed income, and everything is so high, you know. I don't mind preparing the produce at all. I like a lot of salads; I like a lot of vegetables."

Because Full Belly Farm is, by design, a farm that seeks to operate profitably within a reasonably local market—growing and selling produce year-round directly to customers in Sacramento and the larger Bay Area—it can easily fill the food bank's request for regular deliveries of produce to multiple drop-off points. At any one time the farm is likely to have up to 50 crops growing.

"For us, the logistics of making those deliveries is part of the value-added that we can provide. Basically, we pick it today, and we have it in those service centers tomorrow, really fresh," said Judith Redmond, co-owner and co-founder of Full Belly Farm. "We have our own trucks, we have our own professional drivers, and we just go out and do it.

As a nation, we've sort of lost our ability to respond locally. We rely on a much more centralized food system, and in that centralized food distribution you lose freshness, you lose connection with the farm, you lose a lot of things."

Deborah Rich is a freelance journalist who lives in Monterey County, California.

Totally Un-COOL:

Why is It So Hard to Find Out Where Your Food Comes From?

Ever wonder about the nationality of those fresh foods in the fridge? Whether the apples or pears were grown in orchards in Washington State, New Zealand, Chile, or China? How about the tomatoes, catfish filets, or salad greens? Believe it or not, a frozen block of ground beef could actually contain a medley of meat and animal parts from 20 different countries.[94]

While labels clearly track the origins of clothing, manufactured goods, and even handicrafts, knowing where your food is grown is another matter entirely. That could change, however, if Congress funds and enforces a rule passed in the 2002 Farm Bill regarding country-of-origin food labeling, or COOL.

After extensive lobbying efforts by consumer and sustainable farming advocates, Congress included COOL legislation under title X (Administration and Miscellaneous) of the 2002 Farm Bill. The rule stipulated that perishable agricultural commodities such as meats, nuts, seafood, and other items be clearly labeled displaying their country of origin by September 2004. There's just one hitch. Under pressure from agribusinesses—who are increasingly shifting livestock or vegetable production overseas to exploit cheap labor or avoid health and environmental regulations—the appropriations committee has repeatedly denied funding for the program.[95] Two years after the mandated deadline, only seafood fell under country-of-origin requirements. Congress delayed implementation of COOL for all other products until September 30, 2008.[96]

The stage is set for the issue to escalate if Congress fails to enact the rule it has already determined is necessary to protect public health. Over recent years, outbreaks of salmonella poisoning have been traced to cantaloupes imported from Mexico; incidences of brain-wasting mad-cow disease, passed on through infected meats, have resulted in hospitalizations and even deaths.[97] Meanwhile, agribusiness giants like Tyson Foods and Cargill have proactively joined forces with the National Cattlemen's Beef Association, the National Pork Producers Council, and other trade groups and hired a high-powered

Washington lobbying firm with the express purpose of quashing labeling rules.[98]

John Tyson, CEO of Tyson foods and an outspoken critic of the country of origin labeling, has protested that the rule would be unduly burdensome and penalize producers and retailers.[99] In fact, the USDA estimated that COOL could cost the food industry as much as $2 billion in paperwork in the first year of implementation.[100] In a separate study, however, the General Accounting Office found USDA's estimate "questionable and not well supported."[101]

The meat industry is concerned that U.S. consumers could become uncomfortable about the safety of meat produced in other countries and therefore less inclined to purchase it. "If you're a large corporation, you're concerned about COOL because it begins to pull the veil back on how you do business," explains Brian Snyder of the Pennsylvania Association for Sustainable Agriculture (PASA).[102]

Large integrated food producers are moving more and more toward overseas food production where workers have fewer rights, earn far less, and where health and environmental regulations are often less stringent.

Reporter Alan Guebert writes that lobbyists from the National Cattlemen's Beef Association, the National Pork Producers Council, and the American Meat Institute have successfully thwarted country-of-origin labeling laws by showering members of the appropriations subcommittee on agriculture and the agriculture committees with contributions.[103]

Despite the David versus Goliath scenario now facing country-of-origin labeling, Snyder and many other sustainable food advocates believe COOL could be included in the next Farm Bill. "People actually care where their food is produced, especially when you have mad-cow disease, avian flu, and other nutritional concerns out there," says Snyder. "The public might react if they see where food is coming from, and this is scary to the corporate farm system".[104]

16. Energy: Farming in an Era of Fossil Fuel Scarcity

No other issue may more profoundly influence future agriculture policy debates than the multi-headed hydra of energy.[105] Energy is the driving force behind our modern food and farming systems, from natural gas-rich nitrogen fertilizers, to electricity for processing and irrigation pumping, to fuels for trucks, tractors, and laser-guided farm equipment. (See Figure 19, *How U.S. Agriculture Uses Energy*, and Figure 20, *How the U.S. Food System Uses Energy*.)

An estimated 20 percent of current U.S. fossil fuel consumption is used to grow, process, and distribute food. Agriculture now contributes 15 percent of total greenhouse gas emissions worldwide, including almost one-quarter of carbon dioxide emissions, two-thirds of methane emissions, and nearly all nitrous oxide emissions.

Energy issues intertwine with other critical concerns. Dependence on diminishing off-shore petroleum resources has finally forced the market to embrace the development of alternatives. As a result, many are calling for Farm Bill subsidies that shift away from cheap raw materials, livestock feeds, and soil conserving set-asides toward growing crops for liquid fuels. There is also the ominous specter of climate change, with the potential to radically alter agriculture as we know it. Climate experts are responding with calls for immediate reductions in air emissions, new levels of energy efficiency, a transition toward organic and perennial agriculture (that reduce energy inputs), and large areas kept out of production to recapture carbon dioxide and buffer against flooding and extreme weather events.

The Oil We Eat

As late as 1910, 27 percent of all U.S. farmland was still devoted to growing feed for horses used in transportation and cultivation.[106] Today, many observers argue that modern farming has become so dependent on fossil fuels that we are literally "eating oil." It's not hard to see their point. Here's just a taste:

- 20 percent of current U.S. fossil fuel consumption is used to grow, process, and distribute food.[107]
- On average, 10 calories of petroleum are needed to yield just one calorie of industrial food (not including transportation).[108]

- Harvesting a single bushel of corn requires two-thirds of a gallon of gasoline.
- The average 1200-pound steer consumes 35 gallons of oil–nearly a barrel–over its short lifetime from cow-calf operation to feedlot.[109]
- Nitrogen fertilizers, synthesized from natural gas, are the backbone of high-yield industrial agriculture, consuming nearly one-third of the energy used in U.S. agriculture.

On top of all this is an ever-lengthening "food mile" component. With export-oriented farm economies and year-round supplies of produce now the norm, food is transported hundreds or even thousands of miles from farm to table.[110] Burning gasoline and diesel sends carbon dioxide into the already overcharged atmosphere. Agriculture now contributes 15 percent of total greenhouse gas emissions worldwide, including almost one-quarter of the carbon dioxide emissions, two-thirds of methane emissions, and nearly all nitrous oxide emissions.[111]

Growing Food, Feed, Fiber, *and Fuel*?

With the tipping point of "Peak Oil" upon us and the costs of oil and off-farm energy inputs soaring, some "clean-energy" advocates are looking to agriculture to shift our dependence from volatile Mideastern oil reserves to liquid "biofuels" and "biomass" energy derived from Midwestern farm fields. After all, early automobile fuels came from vegetable oils, as did the first plastics. Biofuels, made from fast-growing plant materials, have earned a reputation as clean burning energy. But before we pump up grain-based and cellulosic ethanol and biodiesel as the "silver bullets" that will miraculously wipe away our liquid fuel problems, a little deeper analysis is warranted.

First is the simple energy in, energy out equation. It is seldom mentioned, for example, that on a per-gallon basis, corn-based ethanol delivers only two-thirds the energy content of gasoline (and therefore just two-thirds the traveling power).[112] Meanwhile the decades-old debate is still raging about the amount of power you actually get out of ethanol for what's required to grow and refine it: the "net energy balance." When all of the "well to wheel" inputs of growing, fertilizing, irrigating, harvesting, drying, and processing are tallied, recent estimates reveal that at least two-thirds of a gallon of oil are needed to produce a gallon equivalent of ethanol (roughly a 30 percent net gain). But longtime researcher David Pimentel of Cornell University claims the opposite: ethanol production results in a 30 percent energy loss.[113] Then there is the electricity needed to dry, process, and refine corn into biofuels. Much of it currently comes from coal- or natural gas-fired power plants.

Depending on which life cycle assessment you read (there are dozens to ponder), the shift from hydrocarbon- to carbohydrate-based fuels could either ease particulate emissions and global warming significantly or actually make things far worse. In 2005,

Figure 22

How U.S. Agriculture Uses Energy

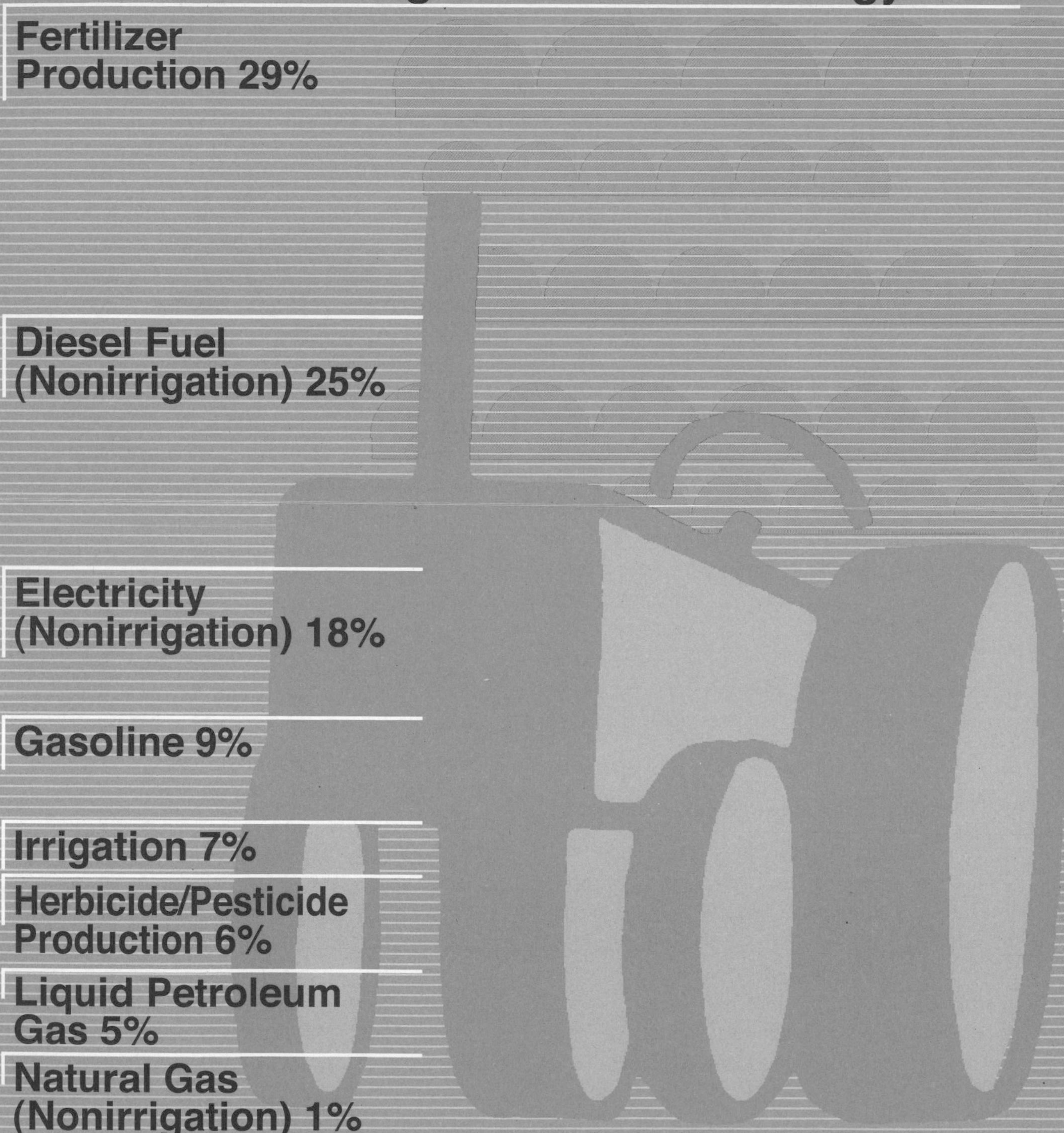

Fertilizer Production 29%

Diesel Fuel (Nonirrigation) 25%

Electricity (Nonirrigation) 18%

Gasoline 9%

Irrigation 7%

Herbicide/Pesticide Production 6%

Liquid Petroleum Gas 5%

Natural Gas (Nonirrigation) 1%

Source: Compiled by Earth Policy Institute from USDA; USDOE; Duffield; Miranowski.

Figure 23

How the U.S. Food System Uses Energy

Home Refrigeration & Preparation 31%

Agricultural Production 21%

Processing 16%

Transportation 14%

Packaging 7%

Restaurants 7%

Food Retail 4%

Source: Heller & Keoleian, "Life-Cycle Based Sustainability Indicators for Assessment of the U.S. Food System," 2000.

Dan Kaman of the University of California at Berkeley's Energy and Resources Group reported a 10 to 15 percent per mile reduction in greenhouse gas emissions from corn-based ethanol. On the same campus, Tad Patzak argues that in its present form, ethanol produces 50 percent more carbon dioxide and sulfur emissions (along with lung and eye irritants) than fossil fuels.[114]

Land Capacity Concerns

Even the most ardent proponents admit that at best, biofuels can only ever be a part of an integrated and diversified energy future. There is simply not enough French fry grease in the world to satisfy the world's diesel addiction, and only so much arable land that already demands careful stewardship. David Morris, founder of the Institute for Local Self-Reliance in Minneapolis and a long-time advocate of a carbohydrate economy writes:

> *The soil cannot satisfy 100 percent, or even a majority of our energy needs. To supply 100 percent of our fuels and electricity we would need 7 billion tons of plant matter, over and above the 1 billion tons Americans already use to feed and clothe ourselves and supply our paper and building materials. Even the land-rich United States lacks sufficient acreage to come close to growing that quantity.*[115]

Using corn as a primary fuel stock, an estimated 25 million acres (39,000 square miles, approximately the size of all the cropland in Iowa or Minnesota) would be converted to corn monocultures just to displace 10 percent of the U.S. oil imports projected for 2025–about a million barrels per day.[116]

Regardless of these significant land requirements, the ethanol industry has grown steadily for 20 years, primarily in the Midwest thanks to generous federal and state subsidies (largely in the form of excise tax exemptions of over 50 cents per gallon, as well as crop supports, production tax credits, and even gas station owner tax relief).[117] (See Figure 22, *The Grain Ethanol Gold Rush.*) By 2005, approximately 15 percent of the U.S. surplus corn harvest was fermented into *gasohol.* That four billion gallons of ethanol contributed about 3 percent of the nation's gasoline–primarily as an oxygenate to replace the toxic MTBE additive.[118]

What began as a movement of farmer-owned and -operated small-scale plants has given way to massive wet mill facilities dominated by global giants like Archer Daniels Midland, one of the most vocal ethanol advocates and most prolific beneficiaries of ethanol subsidies.[119] According to the Renewable Fuels Association, a $2.4 billion ethanol-plant construction boom is now underway. Spurred on by the 2005 Energy Policy Act's lucrative production incentives and mandate of 7.5 billion gallons by 2012, engineering firms like Broin and ICM as well as locally owned cooperatives have joined the grain-based ethanol gold rush.[120] Dozens of relatively small "dry grind" plants (15 to 30 million gallons per year) are being erected in proximity to large grain sup-

Figure 24

Bio-based Energy: Pros and Cons

Potential Benefits	Challenges
Displaces foreign oil. As oil prices rise and supplies contract, a transition beyond a petroleum-based economy is necessary.	Diverts us from a real need—conservation and energy reduction—by reinforcing the "drive forever" and sprawl mentality.
Helps in the transition to new generations of more efficient types of biofuels, such as cellulosic ethanol.	"Net energy balance" of grain-based ethanol not worth marginal fuel production. ("Cold hydrolysis" of starch may begin to offset these energy demands.)
Could potentially shift surplus corn-based agriculture to more diverse farming systems that may have less impacts, such as perennial crops like switch grass.	Still dependent on industrial monocultures that impact a heavy burden in soil loss, toxic inputs, contamination of waterways, loss of habitat, draining of aquifers, and so on.
Reduces greenhouse gas emissions.	Doesn't drastically improve air quality and burns only two-thirds as effectively as gas; corn-based ethanol production has large energy and water requirements.
Offers farmers an additional market for surplus grains. New smaller dry-grind plants could place control of ethanol plants in "farmers' hands."	Early farmer-owned cooperative model is giving way to large-scale shareholder-funded industry with huge subsidized profits at stake.
Helps with future development of crop "residues" such as straw, stalks, and other by-products for primary fuel source.	Puts increasing pressure on farmlands, forests, and native habitats to produce energy in addition to food, feed, and fiber.
Biotechnology can overcome obstacles with specifically designed energy crops, innovative enzymes, and other breakthroughs.	Biotechnology's unacknowledged impacts include uncontrollable cross-pollination, the creation of resistant weeds and organisms, human health allergies, and corporate concentration of wealth and seed supply.
Biofuels are just part of a larger integrated future energy strategy.	The land requirements are burdensome. Replacing just 10 percent of the U.S. liquid fuel supply would require 25 million acres of biomass or an area equivalent to the entire cropland in Iowa or Minnesota.
Provides farmers with new markets and opportunities for farmer-owned multinational cooperatives.	Perpetuates dependence on massive-scale feedlot operations that use grain by-products for feed and on large corporate conglomerates.
Transportation/handling infrastructure could boost local economies.	Crop-based ethanol has limitations in terms of seasonal availability, vulnerable storage, and transportation.

plies, thanks in no small part to crews of temporary Mexican laborers who take on the dangerous task of building the towering concrete silos to house the fuel stock.[121] Meanwhile, farm states like Minnesota and Nebraska are adopting ethanol friendly policies to drastically increase the amount of fuel their farmers grow.

Food Crop versus Cellulosic Ethanol

Ethanol can be divided into two classes: *food-crop* and *cellulosic*. Food crop or "starch-based" processing uses enzymes to break down corn, sorghum, and even wheat gluten into simple sugars, which are later fermented into ethanol. As one of 50 different products generated in a giant wet-mill processing facility (where high-fructose corn syrup and other processed food ingredients are made), this is an energy- and water-intensive endeavor. Dry-grind mills, on the other hand, mechanically turn corn into a fuel stock along with distiller grains and solubles, a by-product fed to livestock. And while dry-grind operations are more energy efficient, even true believers in the biofuel movement question the ultimate resourcefulness of food-crop and starch-based processing. At most, the grain-based ethanol boom is perceived as an intermediate step toward higher levels of efficiencies.

Cellulosic ethanol transforms cellulose–the fibrous leftover stalks of food crops (corn stover and wheat straw for instance), woody materials from fast growing trees and shrubs, stems from grasses, even paper and cardboard fibers from the municipal waste stream–into a fuel stock. Cellulose can be chemically processed into a "syngas," or fermented and broken down with enzymes.[122] A huge task lies ahead, however, if cellulosic ethanol will ever become an economically practical North American liquid fuel source. Writing in the *Wall Street Journal*, John Deutch, a former energy official in the Carter Administration, explains:

> *Biotech experts need to assemble the "gene cassette" and the organisms, and talented engineers need to demonstrate a cost-effective process. Most importantly, an integrated bioengineering effort is needed to develop a process that reduces the harsh pretreatment required to dissolve the solid cellulosic feedstock, increases the concentration of ethanol that is tolerated by the enzymes, and achieves an efficient process to separate the ethanol from the product liquor.*[123]

With immense tracts of tropical rainforest converted to intensive sugarcane production, an ample water supply, and ten years of government subsidies, Brazil has become a leading producer of ethanol. (By 2006, approximately 40 percent of Brazil's fuel came from sugarcane, yet only a fraction came from cellulosic processing. The country is now poised to become an ethanol exporter as gas prices rise.) North American cellulosic research lags even further behind. Iogen, a Canadian company that has received support from both the Canadian government and Shell Oil, appears to be the closest to commercial viability of

Figure 25

The Grain Ethanol Gold Rush

U.S. Corn-based Ethanol Production 1995-2008

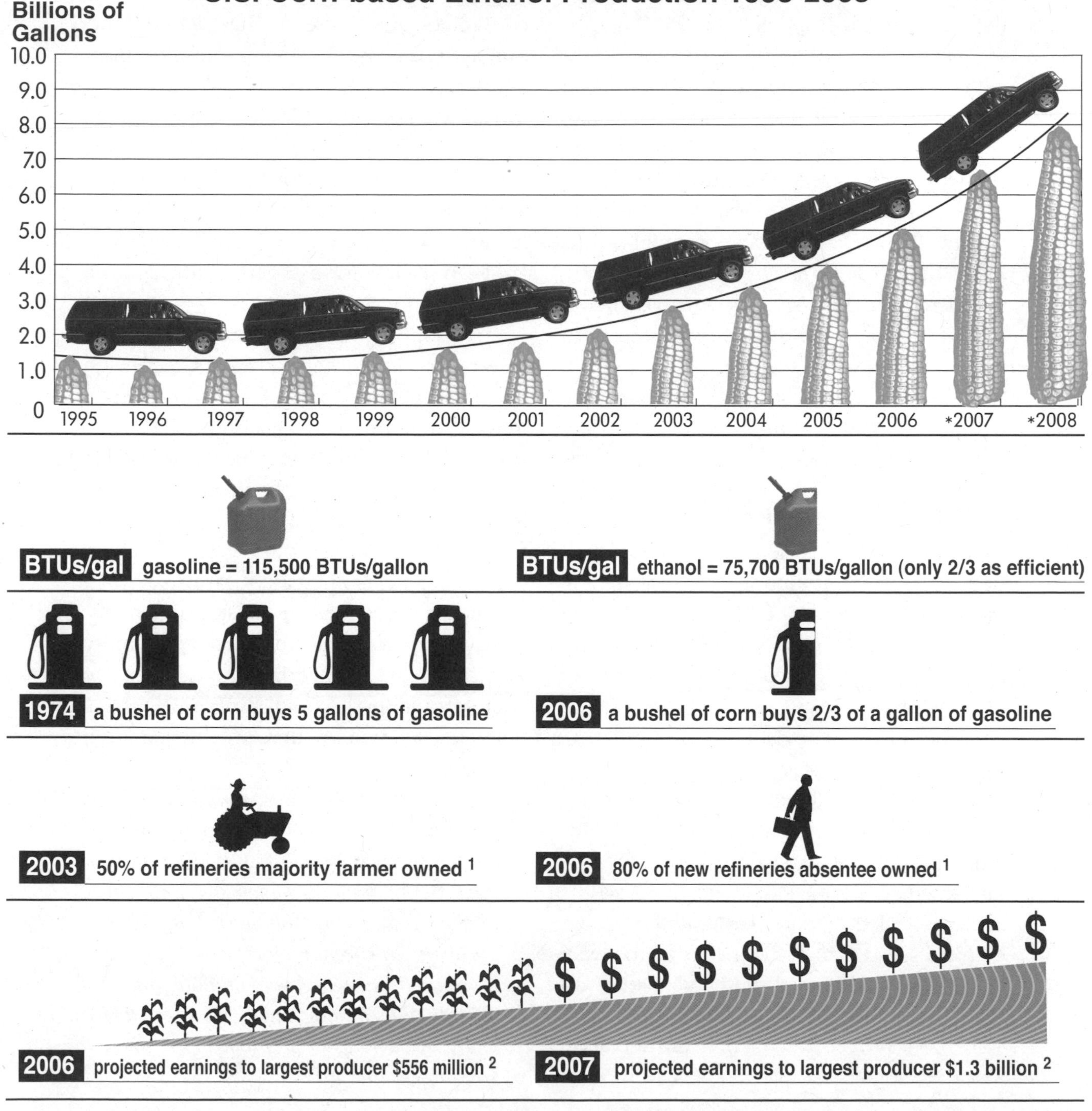

1. Institute for Local Self-Reliance. 2. Alexei Barrionuevo, *New York Times*, Sunday, June 25, 2006.

Sources: The American Coalition for Ethanol and the Renewable Fuels Association. *Projected amount of ethanol to be produced in 2007 and 2008.

cellulosic ethanol production. Experts predict they may still be years away.

The holy grail for a North American cellulosic ethanol industry is predicated on a massive shift from annual crops like corn, sorghum, and soybeans toward permanent or long-term plantings of native plants, such as switch-grass, or forest "thinnings" or high-biomass perennial crops like Chinese *myscanthus* that theoretically won't require excessive plowing or chemicals to pump up yields and wage war with pests and weeds. It's a compelling notion. Still scientists and researchers remain concerned about the ultimate impacts of any intensive cultivation.

Ethanol's Stewardship Legacy?

The drumbeat for a heavily subsidized biofuels industry is mounting. A national campaign named "25 x 25" is aimed at producing 25 percent of the country's fuel consumption by 2025. DuPont has pledged to obtain 10 percent of its energy and 25 percent of its feedstocks from renewable sources by 2010. The "Billion Ton Study," conducted by the USDA and Oak Ridge National Laboratory estimates that the equivalent of current OPEC imports could be replaced by corn and cellulosic ethanol by 2030.[124] Massive set-aside acreages called "Biofuel Crop Reserves" are being discussed and mapped out the way leaders once addressed the need to idle vast highly erodible lands for wildlife habitat and soil protection through the Conservation Reserve Program (CRP). Some family farm advocates are calling for a government-funded biofuel grain reserve that would function as an income-generating transition from corn subsidies. And as the din for more acreage dedicated to biofuels grows louder, many conservationists understandably worry about the vulnerability of transforming every acre–even Conservation Reserve Program set asides and protected grasslands and parklands–into some form of genetically identical biofuel monoculture.

In 2006, World Resources Institute researchers Liz Marshall and Suzie Greenhalgh examined the economic and environmental impacts of a seemingly realistic rise in ethanol production, from the current output of 5 billion gallons per year to 15 billion gallons. Such a jump in ethanol production, according to their report, could generate many positive benefits–increased farm revenue, significant reductions in subsidy payments, lower greenhouse gas emissions, and a more diversified fuel supply. But these gains, the authors cautioned, could also come with a hefty–and potentially unaffordable–environmental price tag.

> *Given current grain-based ethanol technology and in the absence of policy intervention, however, these benefits will come at a cost to our nation's water and soil health. An expanded ethanol market is likely to provide an incentive for farmers to revert to intensively managed rotations and less sustainable management practices, which may have long-term implications for soil and water quality.*[125]

As this book goes to press, standards are being proposed for cultivating and harvesting crops that would provide the raw materials for this new energy infrastructure. Definitions for farming practices that protect the long-term health of the land are an essential starting place. But we can't stop there. Viewing crop residues such as corn stover and wheat straw as merely waste material vastly discounts their value in being returned to the soil as organic matter. Developing ways to boost soil productivity through cover cropping, crop rotations, and other natural methods will become essential as both the economic and environmental costs of petroleum-based fertilizers continue to rise. In addition, harvesting fiber from lands now set aside to protect wildlife could have devastating consequences to biodiversity and could potentially reverse decades of gains made by Farm Bill conservation program efforts. The desirable scale of processing operations, must also be addressed through standards that may affect public financing initiatives. Here are just a few guiding questions:

- Will the introduction of "biofuel crop reserves" impact idled private lands that are now protected for conservation?
- Will parks, forests, and other state and federal lands become vulnerable to energy exploitation and food production?
- How will the challenges of seasonality, storage, and transport of crops and biomass fit into the economic models of biorefineries?
- Will the massive expansion of an ethanol industry be based on monocultures, genetically engineered crops, and chemically intensive agricultural practices?
- Will biorefineries continue to be linked to the confinement feedlot operation model (with feed as a money-making by-product)?
- Will the energy crisis result in a food production crisis?
- Will forests be burned in order to make way for fuel plantations as is now happening the Amazon and Southeast Asia?
- Can subsidies be structured to protect farmers during price falls, and to protect taxpayers from huge payouts to biofuels producers that no longer need them?[126]
- What about other sources of cellulose, such as the Municipal Waste Stream or forest thinning, that could complement and relieve some of the pressure from immense agricultural expansion?

Perhaps there ultimately will be a long-term benefit in all of this, once ethanol ceases to be a profitable way for huge corporations to dump excess corn, and a more logical energy order arises. Perhaps at some point, the biofuel movement could evolve around a diversity of fuel and nonfuel crops on integrated landscapes that include crop rotations, streamside protection, the maintenance of healthy soils, and abundant wildlife habitat and wild areas.

Figure 26

The Answer Is . . .

Integrated Energy Solutions

National Policy on Fossel Fuels and Global Warming

CONSERVATION
Behavioral adjustments and efficiences throughout all levels of society; (Nega-Watts)

BIOFUELS
A commitment to cellulosic technogies and urban/rural materials stream integration

TRANSPORTATION EFFICIENCY STANDARDS
High MPG autos and revived public rail system

SMART GROWTH
Regional priorities for both urban and rural development and protection

RESEARCH AND DEVELOPMENT
New complex crop integration and cover crops to replace fertilizers

RENEWABLE ALTERNATIVES
Farm-scale and appropriate utility-scale renewables; consideration of carbon tax

REGIONAL FOOD CHAINS
Building networks of producers and consumer communities region by region

Thoughts on Acceptable Wind Turbine Placement

The scientific consensus on global climate change predicts inevitable disruptions and potentially dire consequences. The prescriptions are equally clear: significant reductions in fossil fuel emissions are being called for across the board. Agriculture is no exception. Tough choices will be made in the decades ahead. Regional production of diverse renewable energy sources should be aggressively scaled up. At the same time, energy is not renewable if essential resources such as soil and water are despoiled in the process. Simply increasing the supply of renewable energy without a national strategy to make the United States "carbon neutral" may only succeed in providing more power to consumers. Across the world, and prominently in agricultural areas, large wind farms are gaining traction as alternative electricity producers. The latest generation of turbines have been criticized as noisy, aesthetically polluting, and being "Cuisinarts for birds," particularly raptors. Within an overall context of a more positive energy future, however, it should be possible to identify appropriate areas to locate utility-scale wind farms with exceptions such as these proposed by John Davis of the Adirondack Council:

- No energy production in roadless areas.
- No windmills or energy production in wildlife migration corridors.
- No windmills in parks or protected areas.
- Keep windmills away from water bodies.
- Complement renewable energy funding with a national energy conservation platform.

The Farm as Power Plant

One of the Farm Bill's greatest strengths has resided in its capacity to serve as an economic catalyst. In 2002, the Renewable Energy and Energy Efficiency Improvements Program was launched to jump-start a "clean energy" initiative within Farm Bill programs. Known as Section 9006 grants, these funds provide cost-share and loan guarantees that invest in on-farm renewable energy systems, promote energy auditing and conservation, and help to diversify energy sources in rural areas. In addition, investments in renewable energy–unlike corn subsidies for ethanol–are recognized as Green Box payments by the World Trade Organization rules that many in Congress have recently publicly endorsed.

Flipping through a catalog of innovative Section 9006 projects published by the Environmental Law and Policy Center in Chicago, it's easy to get excited about taxpayer dollars being directed toward projects that help farm and food operations become more energy efficient and energy self-reliant. The idea behind these and other projects is to "clean up" the energy flow within farm operations using local solutions and to spill over any excess power to regional markets. Farms are receiving funds to cover rooftops with photovoltaic panels for electricity generation. Solar water collectors are being installed to heat water with the rays of sun. Nuts and fruit crops are being dried in new solar drying systems. Section 9006 grants are also partially funding wind turbine installation–both farm-scale and utility scale–in gusty areas across the Midwest. Energy efficiency efforts have also been awarded with cost-share grants to upgrade refrigeration equipment and grain dryers and conduct usage audits.

Still, as with many Farm Bill programs, the largest grants presently seem to flow downhill to the largest operators, although many recipients are family owned farms or cooperatives. Feedlots with anywhere from hundreds to more than 10,000 animals are receiving funds to install "anaerobic digesters" that convert manure into methane biogas that can be burned to power boilers. Utility-scale wind turbines are receiving cost-share dollars and loan guarantees as an investment in both cleaner regional sources of electricity and renewed development in rural areas.

Outside of the Energy title, on-farm energy conservation incentives were also integrated within the heavily underfunded 2002 Conservation Security Program (CSP). The few farmers lucky enough to be within the chosen watersheds and awarded CSP contracts could receive financial incentives for:

(1) performing an energy audit;
(2) reducing net energy use by decreasing tillage and the use of fertilizers and other fossil fuel inputs;
(3) recycling used motor oil;
(4) purchasing ethanol and biodiesel;
(5) generating solar, wind, hydroelectric, geothermal, or methane power.

Unfortunately, the financial urgency for expanding these programs has not caught up with the political rhetoric. Senator

Figure 27

Demand Outpaces Supply

Applications versus Funds for Section 9006 Renewable Energy Grants

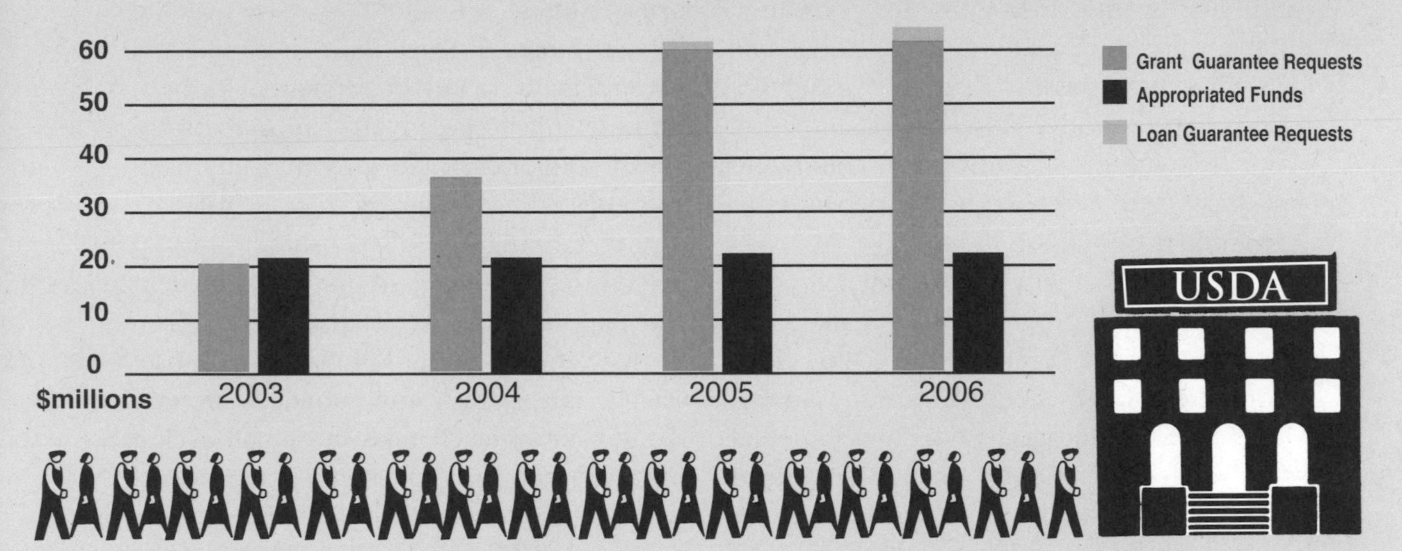

Source: Environmental Law and Policy Center

Renewable Priorities

Distribution of Section 9006 Grants by Technology

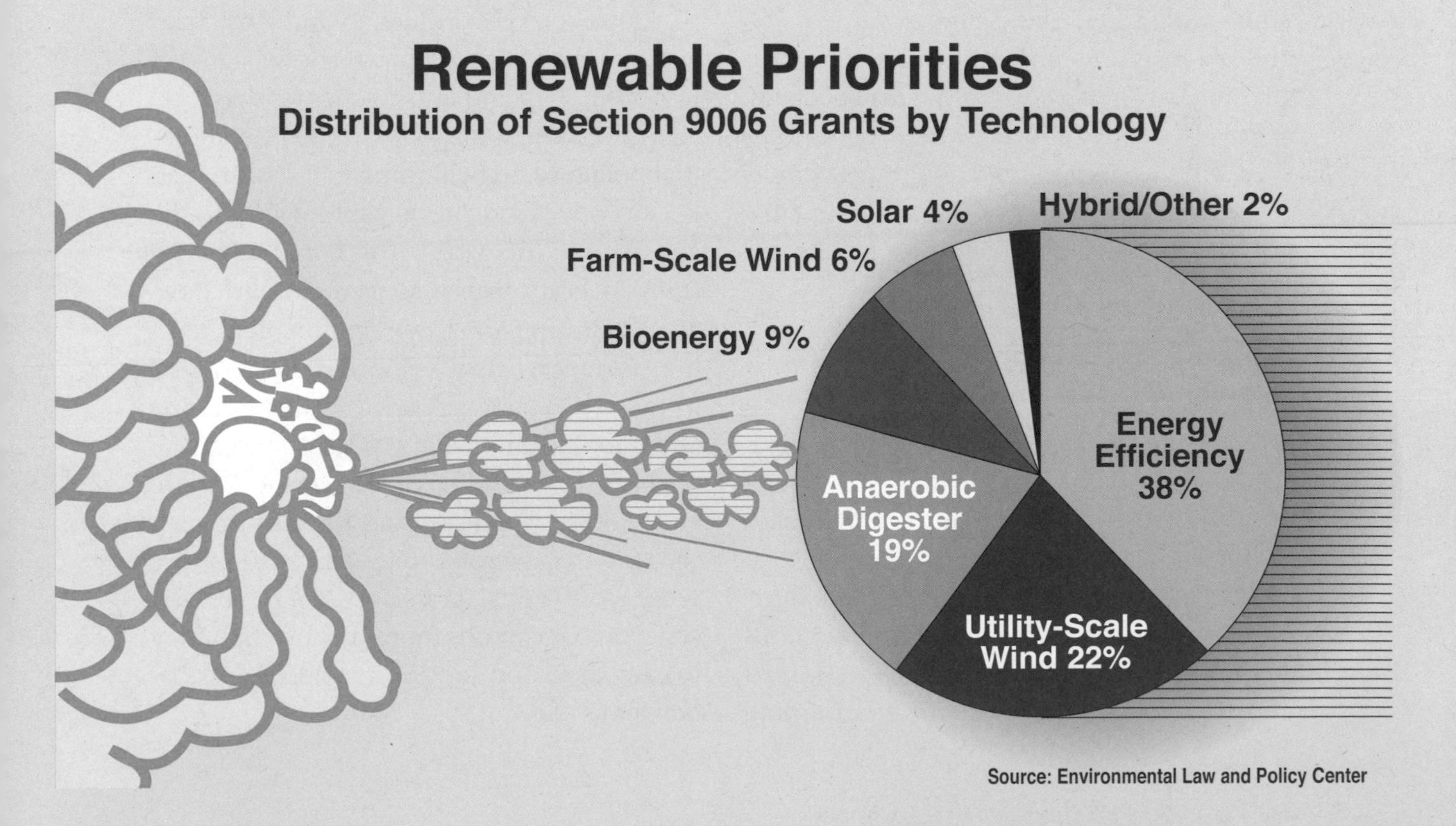

Source: Environmental Law and Policy Center

Richard Lugar of Indiana praised the Farm Bill Energy title, decrying that "our insatiable appetite for energy...represents one of our gravest security threats."[127] Likewise, Congressman Gil Gutknecht of Minnesota was hopeful that such programs could help address "our lack of energy dependence and reliance on fossil fuels."

But like so many environmentally oriented Farm Bill programs, sustainable energy budgets have received paltry allocations compared with other priorities. Groups such as the Chicago-based Environmental Law and Policy Center have been forced to wage hard-fought campaigns every year to defend the appropriations of the Section 9006 annual program budget. Between 2003 and 2005, the USDA awarded $66 million in grants and $10 million in loan guarantees, but they also turned away nearly two out of every three applicants. (See Figure 27 *Demand Outpaces Supply* and *Renewable Priorities.*) Meanwhile ethanol profits, subsidies, and tax breaks are soaring.

Carbon Sequestration—Pollution Off-sets

Perhaps the end-game in energy conservation is carbon capture. This is the natural process whereby living plant material takes carbon dioxide out of the air and ties it back up in the soil in the form of organic matter. It's nature's way of off-setting the tremendous amounts of carbon dioxide (CO_2) pumped into the atmosphere by fossil fuel burning and energy production. Farms with perennial pastures, woodlots, forests, wildlife habitat, reduced tillage, and increased soil cover actually store (or sequester) carbon. Farming practices can also impact carbon emissions. For example, a recent study from various countries shows that small-scale organic farming systems consume 30 to 70 percent less energy per unit of land than conventional farming systems.[128]

As yet, the Farm Bill has no Climate Change title. Few, if any, programs are currently tailored to changes in rainfall cycles, sea levels, air and water temperatures, and vegetation patterns, which scientific consensus insists will inevitably reshape agriculture and life as we know it.

However, governments and the marketplace are beginning to show signs of response. "Cap and trade" schemes that regulate government-mandated limits on industrial polluters are beginning to establish market incentives for carbon capture around the country and the world. This could influence shifts in both crop management and stewardship incentives. Large confinement feeding operations may some day be required to off-set methane gas releases (from farting and pooping animals) by purchasing carbon credits in the absence of a means to capture and prevent its release into the atmosphere, such as an anaerobic digester. Landowners with woodlands, grasslands, and farming systems that capture more carbon than they release may one day be rewarded for that benefit as well.

Conclusion

Energy issues are hitched to a variety of other concerns, as with almost everything related to the Farm Bill. Renewable energy alternatives could save farmers money in the long term. They also represent the kinds of green payments and investments so many are calling for as they look to a whole new era of rural development funded by the Farm Bill legislation programs.

While utility-scale turbines are being erected across the Midwest, critics are arguing against their aesthetic impacts and negative effects on wildlife, and the nuclear industry is lobbying hard to reinvigorate an energy sector that has been stalled for decades. Likewise, as the industrial livestock industry braces for a rise in feed prices due to the biofuel boom, conservationists fear that two decades of Farm Bill protection measures may be reversed to provide domestically produced fuel for gas-guzzling automobiles.

Perhaps even more essential than recognizing the Farm Bill's importance is recognizing its interconnectedness with broader issues and policies. Energy supports of all kinds can become loan-based safety nets rather than long-term entitlements. They can serve broader communities as well as targeted landscape goals, rather than merely profiting absentee plant owners already thriving in the marketplace. We should, not, however, be willing to destroy our farmlands in order to perpetuate the 20th century drive forever mentality. Biological diversity, effects on soil and water, regional economic development, and aesthetics, should all add to the picture of the energy future we are envisioning and investing in with taxpayer resources and the public trust.

GMO Designer Fuels: Coming to a Station Near You

Genetically modified (GMO) fuel crops are coming to a gas station near you. In fact, they might already be in your tank. As of 2004, nearly 45 percent of the U.S. corn crop and upward of 80 percent of all soybeans were planted with genetically modified varieties, primarily approved for animal feed rather than direct human consumption. With consumers in Europe, Japan, Mexico, and Africa increasingly reluctant to allow surplus U.S. RoundUp Ready, Bt, Star Link, and other GMO crops inside their borders, the rapidly expanding North American biofuels industry is set to become a convenient outlet.

In this most recent phase of the grain-ethanol boom, rival seed giants Monsanto and DuPont are jockeying for market share with conventionally bred corn varieties that boast higher starch content to maximize ethanol production. But most experts acknowledge that using corn as a fuel source has its limitations (and negative ecological implications), and some are turning to biotechnology for alternatives.

"More miles to the acre" may be the new mantra of biotech agribusiness firms eager to cash in on the biofuel craze in at least two different ways: (1) modifying the genetic structure of plants to make fermentation easier; (2) boosting yields of both annual and perennial crops. Syngenta, for example, has its sights set on "self-processing" corn. Each transgenic kernel would carry an amylase enzyme that is currently added separately to starch at the ethanol plant.[129] To pull this off, engineers have had to reach outside the corn genome and insert a gene from a thermotrophic microbe that lives near hot-water vents on the ocean floor.[130] Meanwhile, DuPont and Bunge have engaged in a joint-venture to genetically engineer soybeans for biodiesel fuel and other uses.[131]

Researchers are also branching out from annual feed and oilseed crops into perennial plants potentially high in biomass—fast-growing poplar trees and dense grasses such as the Chinese *myscanthus*, which promoters tout can grow with next to no inputs or irrigation and produce 20 tons of plant materials per acre. California-based Ceres Corporation has been conventionally breeding switch grass, a Prairie states native, which it claims would need less fertilizer and irrigation and require infrequent replanting.

In addition to boosting yields, another primary goal of biotech firms is to reduce the amount of lignin that holds plant cells together. Removing lignin presently poses a complicated step in converting cellulose into ethanol. But it's also Nature's way of endowing plants with the stiffness to grow upright. Needless to say, the prospect of unleashing a new genetic trait such as "droopiness" onto the landscape is raising the hackles of scientists, botanists, legal activists, and other observers.

Even the conventional agriculture community is cautioning against an escalation in intensive farming practices stimulated by a massive increase in corn production. Many fear that the abandonment of crop rotations could strain the soil, tax water resources, and lead to a buildup of insects or vulnerability to disease.

Figure 28

Who's Growing GMOs

Percent of Global Land Area Planted in Biotechnology Varieties by Country (2003 Total Global Land Area: 167.2 Million Acres)

***Other countries include: Australia, Mexico, Romania, Bulgaria, Spain, Germany, Uruguay, Indonesia, India, Columbia, Honduras, and the Philippines.**

Source: International Service for the Acquisition of Agri-biotech Applications (ISAAA) Global Review of Transgenic Crops 2003.

At the same time, the prospects of new and relatively untested GMO energy crops hold their own set of complications. The transfer of exotic genes and enzymes from energy crops to the human food supply or to the wild are both entirely possible. Pollen transfer in open fields between the same types of crops or their wild plant relatives is a naturally occurring and uncontrollable phenomenon. The intermingling of seed is also almost impossible to prevent. This has been soundly proven with StarLink corn (approved only for animal consumption, it has surfaced in tortilla chips), and more recently, herbicide-resistant Liberty Link rice (which has resulted in plummeting sales for U.S. farmers). Such contamination—of plants and seed banks—are essentially irreversible. There are health implications as well. Bill Freese of the Center for Food Safety in Washington D.C. reports that some amylase can induce allergy and requires further study.[132]

The biotech community seems poised to go to great lengths to produce fuel from agriculture. It's important to determine whether a cultural addiction to liquid fuels and automobiles could possibly be worth such risks.

The issuing of patents on the genetic make-up of plants concentrated in the hands of just a relatively few global corporations is perhaps what most concerns Dave Henson, an expert on genetically modified crops. "GMO biofuel conglomerates have the potential to become the next OPEC," cautions Henson. "Controlling patents and the seed supply means these giants are no longer just grain brokers and dealers, but will have the power to exercise control over growers and communities all over the world."

17. Healthy Lands, Healthy People: Why Farmlands Matter to Conservation

America's conservation landscape is a complicated patchwork, often stitched together or pulled apart by the country's one billion acres of agricultural lands. Forest, pasture, range, and crop lands make up nearly two-thirds of the country's contiguous landmass (see Figure 29 *Land/Cover Use.*) For many threatened and endangered species that have no concept of public and private property lines–from resident and migratory birds, to fish and amphibians, to native pollinators and large roaming mammals–those field margins, grasslands, woodlands, and waterways also serve as habitat and safe passage.

Conservation on private lands remains essential to maintain a network of healthy habitats in regions throughout the country. And the Farm Bill's Conservation Title potentially provides not just millions, but billions of dollars–when appropriated–to accomplish this.

Across the Corn Belt, grassland bird species have experienced a precipitous decline in recent decades. Once-common species such as bobolinks, savannah sparrows, and eastern and western meadowlarks have seen their breeding grounds disappear and their populations plummet as grasslands are converted to subsidized row crops and subdivisions. While some farmers and ranchers have adopted stewardship practices to accommodate these declining species, the critical mass needed to stem the losses has not materialized. Ornithologists warn that we have only a decade to restore perennial ground cover across what is now the corn and soybean landscape before many grassland birds disappear from their traditional ranges altogether.

In the Klamath Basin, which spans the border between southwestern Oregon and northwestern California, the fight over water for irrigation and in-stream flow for dwindling fish populations has escalated to a boiling point. Farmers are clinging tenaciously to historical water rights and subsidy programs to keep their operations going. So far, they are winning. In the mean time, native fish–including steelhead, green sturgeon, five species of salmon, the Lost River sucker and shortnose sucker, and two species of mullet–are on their last gasps. Without extensive government supports, farming in the Klamath Basin would be unfeasible. The region's chief crops–onions, potatoes, and grains–are in oversupply. They're even sometimes plowed under when prices fall too low. The trouble is, the Klamath region's water wars are just the start of similar struggles between agriculture, wildlife protection, and development simmering all over the country and throughout the world.

Paying Farmers Not to Farm

Why pay farmers not to farm? It's a natural enough question. The answer partly lies in the values and benefits that healthy rural landscapes can provide that are not compensated by the almighty marketplace. Water filtration and flood prevention, open space preservation, wildlife habitat, carbon storage, scenery enhancement, and species protection are just a few of the multidimensional aspects of agriculture that can occur on well-cared-for private lands. But farmers forced to maximize profits through intensive production often work directly against these benefits: by diminishing biodiversity with genetically identical monocultures, mining the soil, overdrawing groundwater reserves, physically concentrating livestock in huge numbers, or continually applying agrochemicals.

Any citizen who cares about land stewardship has a stake in what happens on farmlands. Here are a few compelling reasons:

- Seventy percent of the U.S. land base is privately owned.
- Nearly two-thirds of the 1.9 billion acres in the continental U.S. is comprised of crop, pasture, range, and forest land–about one-half of which is privately owned.
- Only one-tenth of the lower 48 states fall under some form of state or federal habitat protection, and these areas have become increasingly fragmented and isolated.
- Public lands are being exploited for resource extraction, grazing, timbering, off-road recreation, and other harmful activities.
- Every year, 1.2 million acres of agricultural and forest lands are lost to development.[133]
- As of 1995, 84 percent of all endangered or threatened plants and animal species were listed in part due to agricultural activities.[134]
- Between 2006 and 2010, nearly 28 million acres under Conservation Reserve Program contracts will expire; their future is uncertain.
- Conservation programs are among the Green Box payments that are acceptable under World Trade Organization rules.

With the U.S. population now at 300 million and showing no signs of a plateau, agricultural lands face constant development pressure. In California alone, an estimated 125,000 acres of irrigated fields, pasture, and other rural lands are developed each year.[135] Nationwide the annual losses reach ten times that number. At the same time, the younger generation is quickly losing its motivation to enter the farming profession. Farmers over 65 years of age outnumber those under 35 by two to one. And over the next few decades, an estimated 400 million acres of agricultural land will transfer to new owners who will

Figure 29

Land Cover/Use

Surface Area of the U.S. Contiguous States, 2002

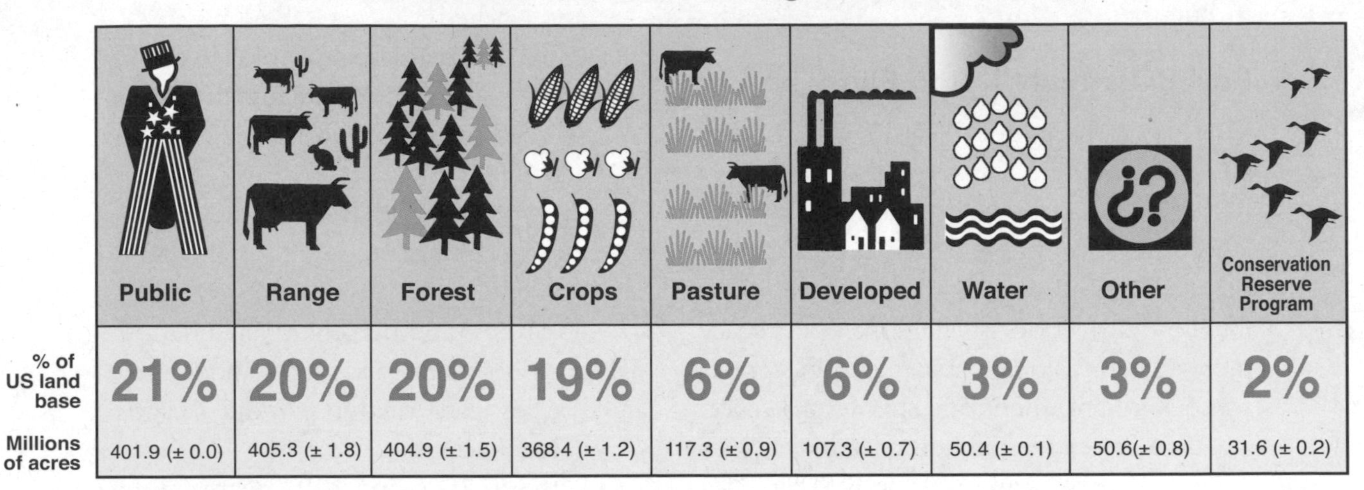

	Public	Range	Forest	Crops	Pasture	Developed	Water	Other	Conservation Reserve Program
% of US land base	21%	20%	20%	19%	6%	6%	3%	3%	2%
Millions of acres	401.9 (± 0.0)	405.3 (± 1.8)	404.9 (± 1.5)	368.4 (± 1.2)	117.3 (± 0.9)	107.3 (± 0.7)	50.4 (± 0.1)	50.6(± 0.8)	31.6 (± 0.2)

Total surface are of the 48 contiguous states by land cover/use in 2002. Margins of error defining the 95% confidence intervals are in parentheses. The total surface area of the United States is 1,937.7 million acres (NRCS 2004).

either continue to farm or convert it to non-agricultural uses.[136]

The use of easements to acquire rural lands (or their development rights) by government agencies and land trust organizations has emerged as an effective tool to buffer against sprawl and protect biodiversity. But the outright purchase of all the land necessary to safeguard our natural heritage is simply not an option. (Even if purchased, many landscapes would require ongoing management, at least in the short-term.) Conservation on private lands, then, remains essential to maintaining a network of healthy habitats in regions throughout the country. And the Farm Bill's Conservation title potentially provides not just millions, but billions of dollars–when appropriated–to accomplish this.

The best programs, says NRCS biologist Randy Gray, take place on a landscape scale. That is, they encompass a range of core habitat large enough to permit the recovery of a community of species. This could include a Montana alfalfa farmer who receives Environmental Quality Incentive Payments for cutting back on irrigation in order to raise the Big Hole River's in-stream flow to support the threatened fluvial Arctic grayling. It could be Utah ranchers in the Parker Mountain Adaptive Resource Management area who earned Wildlife Habitat Incentive Payments in return for changing grazing practices and installing fences so that a quarter million acres of sage grouse habitat can be restored. Or it could be a Lower Mississippi Valley farmer whose Conservation Reserve Program or Wetlands Reserve Program contract has helped to return more than 600,000 acres of agricultural lands to bottomland hardwood forest habitat for the Louisiana black bear and hopefully the Ivory billed woodpecker.

Alphabet Soup: Deconstructing the Conservation Program Palette

Understanding Farm Bill conservation programs requires delving into a parallel universe of acronyms. First and foremost are the Natural Resources Conservation Service and Farm Service Agency–the NRCS and FSA–the conservation arm of the U.S. Department of Agriculture that administers programs on a number of basic levels.

Conservation and stewardship. Farmers and ranchers are stewards of more than 50 percent of the lands in the contiguous United States. We simply can't purchase or expropriate all the land needed to protect species and habitats. This is not limited to soils and water resources but also includes the stewardship of woodlots, forests, wetlands, grasslands, and complex habitat linkages. Economic incentives and assistance for private landowners are critical.

Set-aside programs such as the 1985 Conservation Reserve Program (CRP), 1990 Wetlands Reserve Program (WRP), and underfunded 2002 Grassland Reserve Program (GRP) pay landowners to take land out of production and restore functional grasslands and wetlands. The most effective programs target large areas of contiguous and high-priority habitat and are either permanent buyouts or long-term (30-year) contracts.

Habitat building programs offer cost-share assistance to restore land and protect declining species, and include the Wildlife Habitat Incentives Program (WHIP), Conservation Reserve Enhancement Program (CREP), and others.

Compliance-oriented programs like the Environmental Quality Incentives Program (EQIP) have more questionable conservation values. They often pay large amounts of money to polluting corporations, such as massive confinement hog and dairy farms, to comply with the Clean Water Act, Clean Air Act, and other regulations that most businesses have to abide on their own.[137] However, to a lesser extent, EQIP has also been used to provide habitat for fluvial Arctic graylings, lesser prairie chickens, sage grouse, and bobwhite quail.

Production and stewardship-oriented incentives such as the Conservation Security Program (CSP) were designed to combine both sustainable farming and long-term care for the land. The CSP encourages and supports conservation on farming and ranching operations of all types in all regions, and comprehensively addresses soil, water, wildlife, energy, and other resources as a basis of healthy agriculture rather than as side issues or through costly remediation.

CRP and WRP: A Modern New Deal

The lessons of the Dust Bowl–a continent literally displaced by drought and economic depression–inspired the early Farm Bill conservation programs. In the following decades, sensible practices such as cover cropping, field rotations, contour strips,

Key Events in Farm Bill Conservation Policy

1935	• Establishment of the Soil Erosion Service
1985	• Conservation Reserve Program • Sod Buster (Highly erodible lands)—plowing up intact grasslands and shrubland habitats should be considered a crime • Swamp Buster (Wetland conservation provisions)—conversion of wetland results in loss of USDA benefits
1990	• Wetlands Reserve Program—restore and protect wetlands and riparian zones • Almost 2 million acres mostly permanent • 50 percent or greater can be adjacent uplands
1996	• EQIP • WHIP—develop and improve fish and wildlife habitat • Cost share program (75 percent or better) • 5–15 year contracts • Farm and Ranchland Protection—easement purchases with partners to maintain land in agriculture and associated use • State technical committees
2002	• $15 billion "committed" to conservation programs • Grassland Reserve Program—grazing lands and biodiversity (2004); sage grouse targeted • Conservation Security Program—annual payments for good stewardship with the highest payments going to address wildlife habitat

windbreaks and hedgerows, influenced a new ethic of soil protection. Those conservation practices also had an ulterior economic motive. Farmers who received subsidies were required to idle a portion of their land to limit overproduction. But during the Get Big or Get Out era, Farm Bill conservation programs took a u-turn. Tile and open-ditch draining became standard management practices of USDA's Agriculture Conservation Program. Every year from the mid-1950s to the mid-1970s, more than 500,000 acres of wetlands were lost to agriculture.[138]

Conservation took another twist during the early 1980s family farm crisis, at a time when foreclosures and even suicides had become all too frequent news items. Concern among biologists about declining North American waterfowl populations, whose breeding grounds fell within farmed areas of the northern prairie states, generated a new wave of reforms. Out of the ashes emerged three provisions in the 1985 Farm Bill that charted a new course for wildlife management on private lands. The Conservation Reserve Program (CRP) paid yearly fees to farmers to idle 40 million acres of critical waterfowl breeding habitat and highly erodible cropland.[139] Two "disincentive" policies–Swamp Buster and Sod Buster–revoked subsidy payments to any farmers who drained wetlands or converted prairies into cropland. Legislators continued this trajectory with the Wetlands Reserve Program (WRP) in the 1990 Farm Bill. This program contracted with landowners to restore and protect formerly converted wetlands through permanent easements. The WRP also made a continental linkage between the breeding grounds of migratory birds in northerly CRP lands and the overwintering wetlands in the south.

Looking back, the concepts (though not necessarily the execution) behind the CRP and WRP can be viewed as nothing short of revolutionary. The idling and restoration of millions of acres of marginal lands and wetlands served as a serious stop-gap measure to prevent the conversion of the entire billion-acre agricultural landscape into an ecological sacrifice zone for the production of protein, carbohydrates, and oil seeds. While far from perfect, the CRP and WRP placed nearly 10 percent of all croplands and more than 2 million acres of wetlands under some form of protection. Among the related quantifiable outcomes:

- Annual soil erosion on croplands fell from 3.1 billion tons per year in the pre-CRP era to 1.8 billion tons in 2001.[140] (See Figure 30, *Erosion on Cropland*.)
- More than 26 million ducks are estimated to have been born between 1992 and 2003 as a direct result of CRP enrollments.[141]
- Wetlands increased annually at a rate of nearly 70,000 acres per year between 1997 and 2003.[142]
- Both CRP and WRP were highly popular with landowners. For most programs, application demand outnumbered funding by nearly three-to-one.

With the dark cloud of energy uncertainty hovering over the agricultural landscape, these conservation gains could quickly change. Between 2007 and 2010, nearly 400,000 CRP contracts on 28 million acres are scheduled to expire.[143] (See Figure 31, *What Will Become of CRP Contracts?*.) Many of these lands are already coming out of retirement as a result of shifting conservation priorities and market forces. According to the Sustainable Agriculture Coalition, trends show that at least 4 million acres–approximately 12 percent–of current CRP set asides will be converted back to agricultural lands. This number could be far greater. Spurred on by escalating state and federal incentives for ethanol and biofuel expansion, millions of acres of CRP lands could return to agricultural production. With the U.S. biofuel industry currently dominated by corn ethanol, and a smaller amount of biodiesel produced from oilseeds, this almost certainly signals a return from CRP protection to row crop production.[144] Even on those CRP contracts that remain in effect, a clamor is mounting to allow the harvest of biomass materials for energy and bioplastics production. The future of the Wetlands Reserve Program may be equally uncertain. Congressional appropriations have recently capped the WRP's annual acreage far below the expansion levels promised in the original 2002 Farm Bill. The decline of both programs could bode poorly for many species.

Getting Conservation Programs on Track

Despite obvious good intentions, the past two decades of ever-increasing Farm Bill

Figure 30

Erosion on Cropland

Conservation programs have reduced but not eliminated erosion on farmland

Source: UDSA NRCS

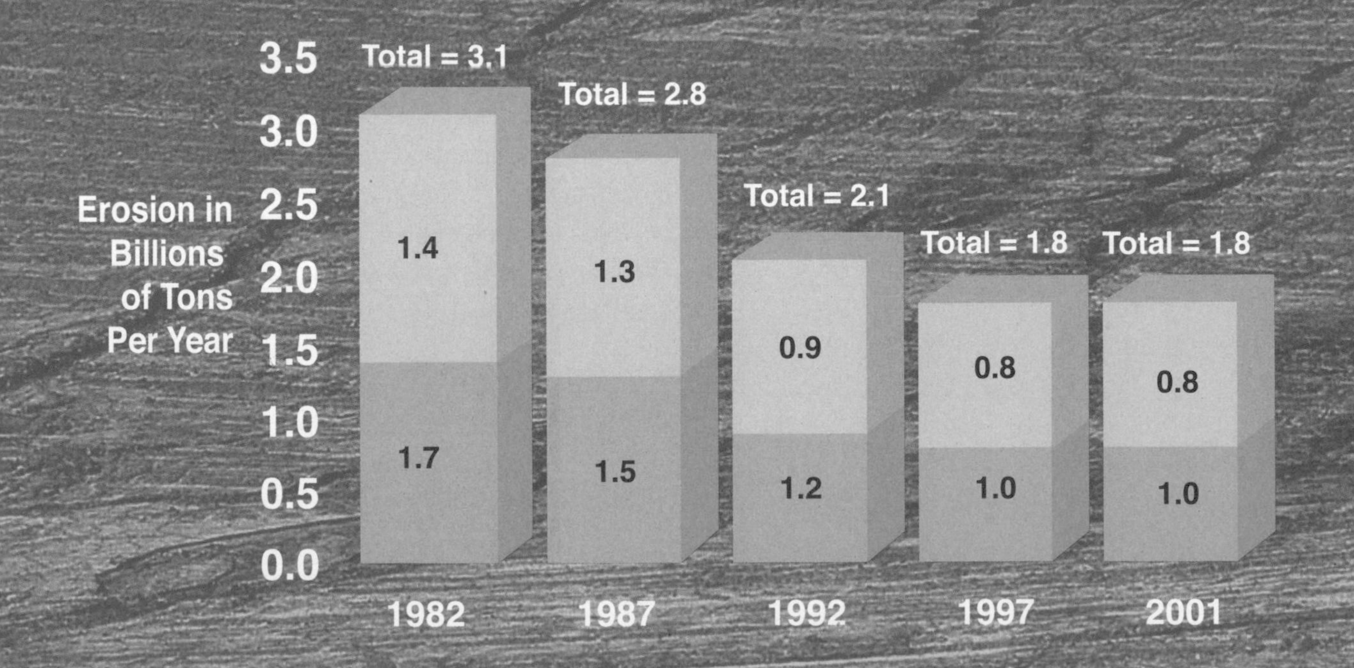

Figure 31

What Will Become of CRP Contracts?

Millions of conservation acres may go under the plow 2004–2015

CRP is at a critical junction with 400,000 CRP contracts on 28 million acres scheduled to expire between 2007 and 2010. USDA has undertaken an intitiatve to re-enroll and extend some of these contracts on a short-term basis. But it remains clear that millions of CRP acres will no longer be idled and instead be brought back into production in the coming decade.

Source: Isaac Walton League

Number of contracts due to expire

180,000
160,000
140,000
120,000
100,000
80,000
60,000
40,000
20,000

2004 2005 2006 2007 2008 2009 2010 2011 2012 2013 2014 2015

conservation spending have also been marked by significant shortcomings. They fall into a few basic categories: (1) inadequate staffing at the Natural Resources Conservation Service (charged with implementing these programs); (2) program eligibility; (3) lack of strategic landscape-based focus; (4) flat-funding.

Early conservation programs were not backed by huge budgets. (Set-asides, in fact, were often required for production subsidies.) Beginning in 1985 and continuing with each successive Farm Bill, however, legislators pushed more and more money at landowners to achieve conservation goals. Meanwhile, budgets for on-the-ground technical assistance have not kept pace with conservation subsidies. In the words of one top official, "NRCS staff have been turned into money obligators, pushing money out the door, frequently to the largest landowners, so that allocations aren't lost by the end of the fiscal year, without always being able to do the conservation planning they have been trained to do." This lack of staffing sends ripple effects across the conservation landscape.[145]

Who receives the money is a second critical problem. Ironically, some of the country's worst stewards have been traditionally rewarded with the most money under the premise that landowners with the most egregious problems deliver the highest benefit per dollar spent. Good stewards, for the most part, have been largely left out of this process. In the worst cases, conservation contracts have inspired opportunistic landowners to plow up and erode intact prairie remnants, or to remove functional terraces and shelterbelts that protected fields and slopes, so they could apply for set-aside payments. In the case of the Wetlands Reserve Program–arguably the Farm Bill's most successful conservation effort to date–only wetlands previously impacted by agricultural development are eligible for funding.

A lack of focus on regional conservation plans has also diluted the potency of conservation programs. Without careful coordination into watershed-wide, habitat-specific, or larger regional and statewide conservation strategies, efforts become diffused across the landscape in a catch-as-catch-can basis. NRCS biologist Randy Gray has described past funding approaches often as "random acts of environmental kindness." According to Gray, to maximize benefits to fish and wildlife, such a scattered funding approach must be replaced by goal-oriented conservation plans that are committed to restoring habitat across agricultural regions, rather than just at the individual farm or ranch level.

By way of positive example, Gray praises the Lower Mississippi Valley Joint Venture as a project that has developed valuable modeling tools for prioritizing conservation spending. "Not every acre of the landscape is created equally," says Charlie Baxter of the U.S. Fish and Wildlife Service, "so it matters where you reforest." To that end, Baxter and his colleagues have developed decision support tools that help them focus on specific biological outcomes across the region.

Senate and House appropriations committees also deserve a heaping share of criticism. Their budgets are normally delivered months late and double-digit percentage points short of the cash promised in the original bill. As their popularity has soared in communities across the country, congressional appropriations committees have consistently flat-funded conservation programs for the past two decades by slashing allocations during annual budget negotiations.

Yet nothing–nothing–continues to be more counterproductive than the complete disconnect between commodity crop subsidies and conservation programs. On the one hand, export subsidies encourage farmers to maximize acreage of basic commodities and have all but eliminated economic risks and encouraged the plowing of even the most marginal lands. Meanwhile, the USDA directs about 10 percent of its overall spending toward conservation, much of that to right past wrongs and to clean up problems related to overproduction. Consider, for example, that while 1.7 million acres were enrolled in the Conservation Reserve Program in South Dakota between 1985 and 1995, during the same period more than 700,000 acres of grassland were converted to cropping. These were grasslands primarily tilled for corn and soybeans, already in excess supply.[146] Such a dichotomy makes essential Farm Bill conservation programs seem more like a distraction than a coordinated national stewardship strategy.

Clearly, conservation is best practiced by rewarding farming and ranching practices that are sustainable in the first place and require no costly mitigation and environmental clean-up dollars afterward. That way agriculture becomes a starting point in a national movement to sustain vital watersheds and ecoregions throughout the country.

An Idea for the Twenty-First Century

In a radical departure from decades of commodity subsidies, the 2002 Farm Bill introduced the Conservation Security Program (CSP), the country's first green payments program. Rather than offering subsidies to maximize commodity output or take land out of production, the CSP rewards landowners for sound stewardship–soil protection, clean water, energy efficiency, and pesticide reduction. Backed by Iowa Senator Tom Harkin, and referred to inside the Beltway as "the Harkin Program," the CSP was crafted to support a whole new era of agriculture, one that could allow farmers to transition away from commodity crop production, increase the equity of subsidy payments, and conform to WTO rules of acceptable agricultural supports.

Like the CRP and WRP, the Conservation Security Program was conceptually brilliant. Designed as a full entitlement program, its

Figure 32

The Conservation Challenge

Ongoing Concerns	Critical Programs and Ideas
Feedlots will continue to be propped up by cheap feeds and grain-based ethanol and gasification expansion	Grass-pastured livestock movement should spread widely, particularly in areas of abundant rainfall
Conservation programs continue to be flat-funded	Conservation title should be fully funded and augmented with more on-the-ground technical staff
Conservation Reserve Program contracts expire and millions of acres may become intensively cropped for biofuel grains	CRP contracts protect erodible land, provide wildlife habitat, and decrease overproduction; energy crops should be produced from diverse perennial crops under sound conservation guidelines
Conservation Security Program remains very slow to roll out across the country	Conservation Security Program has the potential to transform the farm support system and at the same time achieve far-reaching conservation and trade goals

budget was to be open-ended and eligible to all landowners who applied and met its high stewardship standards. With the motto of "rewarding the best and motivating the rest," the CSP was set up to support farmers and ranchers already engaged in conservation with a record of protecting the environment. Successful applicants must demonstrate that they are preventing manure or other fertilizer from running into streams, and that they are conserving soil and minimizing pesticide use, among other requirements. Extra points–and higher payments–are available for additional efforts to provide habitat for wildlife or protect streams and groundwater, including reducing fertilizer or pesticide use, converting crop land into permanent pasture, or installing farm-scale windmills or solar photovoltaic arrays to supply the farm with energy.

Advocates of healthy agriculture agree, hands down, that the Conservation Security Program (CSP) is the most inventive idea to grace a Farm Bill in decades. For the first time since the New Deal era, both conservation and production standards were rolled into a single program. Rather than encouraging damaging high-output commodity agriculture with one title, and funding remedial conservation with another, the CSP embodied a holistic approach, rooted both in sound farming and exemplary stewardship. For the first time, organic farmers–now the fastest growing segment of the food sector and long ignored by Farm Bill spending–had an advantage in program eligibility. The CSP also held the promise of serving as a safety net for farmers interested in transitioning from commodity row crops toward perennial grass pastured agriculture, an urgent reform required of the food and farming system.

Unfortunately, the funding has not matched the expectations promised in the original bill. After a two-year delay, the CSP enrolled nearly 20,000 farmers in 280 selected watersheds. Far from a nationwide entitlement program, just 10 percent of the country's 20,000 watersheds have been exposed to the program benefits.

Healthy Lands, Healthy Economies

From a taxpayer perspective, it's hard to argue against most of the potential benefits of the Farm Bill's conservation dollars. Maintaining wildlife habitats, preserving open space, reducing erosion, and providing clean water are all public goods that are directly dependent on thoughtful stewardship on private lands. Yet we can get better at asking the key questions that put conservation dollars to best effect. *Where are declining species? What are the priority species? Where in the landscape can we get the best results? Should taxpayers compensate farmers many times over the value of their property for not farming (through short-term contracts), or should targeted grasslands, forestlands, wetlands, and waterway corridors be purchased outright?* There are also increasingly convincing economic arguments not just for increasing the Farm Bill's conservation programs, budgets, and on-the-ground technical staffs, but also to make conservation and the preservation of healthy landscapes the basis of a new agricultural support system. In fact, signs show that the more species we keep alive in the places where we live and farm, the more successful our agricultural operations become.

Consider, for example, that one out of every four mouthfuls of the foods and beverages we consume depend on pollination. Farmers grow more than one hundred crop plants that rely on pollinators–bees, butterflies, moths, and even hummingbirds and bats–from apples and cherries to squashes and blueberries. According to the Xerces Society, insect-pollinated crops contributed an estimated $20 billion to the U.S. economy in 2000. If this calculation were expanded to indirect products, such as milk and beef cattle fed on alfalfa, pollinators would be responsible for almost $40 billion worth of agricultural products each year.[147]

Due to a number of environmental factors, the European honeybee, the world's most relied on agricultural pollinator, has declined by 50 percent since World War II. Likewise, the continent's thousands of native pollinators have suffered from the fragmentation of habitats and the extensive use of pesticides. A growing body of evidence supports the restoration of habitats in and around farmlands to allow native pollinator populations to rebound–if only as an insurance policy against predicted catastrophic losses of honeybees. Native habitats in and around farms can support dozens of resident pollinators (as well as other beneficial insects) that eagerly go to work in farm fields and orchards.

Agriculture may not always be the best economic investment in areas of limited water resources, either. This is yet another reason to factor conservation not just as a side issue, but as a diversified component of a sound food, farm, and rural lands policy. According to a peer-reviewed U.S. Geological Services survey, for example, Klamath Basin farmers annually generate $100 million in revenues. Recreation–boating, rafting, camping, swimming, and fishing–on the other hand, brings in $800 million. The report further states that restoring water to the Klamath River could bring a boon of $3 billion to the region's economy. Buying out farms from willing sellers and protecting the land could cost $5 billion, but would produce a 7-fold return of $36 billion.[148]

Studies also show that every dollar invested in riparian vegetation (which helps to filter water and recharge groundwater) saves anywhere from $7.50 to $200 in municipal water treatment. California alone has some 17,000 miles of irrigation canals, mostly void of vegetation and kept bare either by mechanical scraping, mowing or herbicide applications. Clearly a nationwide campaign to improve habitats throughout all of the nation's waterways could have positive impacts on health, wildlife, and regional economies. In fact, Farm Bill dollars are already at work to revegetate thousands of miles of farmland waterways throughout Pennsylvania's and Maryland's tributaries to the Chesapeake Bay.

There is another fundamental reason to further bolster and refine conservation spending. Species, once lost, are gone forever and the fibers of the continent's distinct biological mosaic begin to unravel. The fate of wild nature is dependent on the food and farming systems of our present and future. But this relationship is codependent. Land health and the health of the people will always be deeply interconnected. There will be no agriculture on completely degraded habitats.

Conservation Program Landscape

Conservation Security Program (CSP—2002). The first comprehensive green payments approach to agricultural subsidies. Also the first conservation program enacted as an entitlement program, with a budget set by the number of farmers deciding to apply and able to meet the rigorous environmental standards required. The Bush Administration flat-funded and capped the CSP far below promised levels.

Conservation Reserve Program (CRP—1985). CRP is a voluntary land retirement program in which landowners sign up for 10- or 15-year contracts, receiving annual rental payments and cost-share assistance to establish resource conserving groundcovers on eligible, mostly highly erodible, farmland. Contracts are expiring on 28 million acres between 2006 and 2010, and biofuels advocates are eyeing CRP land for grain and cellulosic ethanol production.

Wetlands Reserve Program (WRP—1990). Helps landowners protect, restore, and enhance wetlands through long-term and permanent conservation easements. The goal is to protect priority wetland functions and values, along with optimum wildlife habitat. The 2002 Farm Bill boosted the WRP to 250,000 new acres each year, but congressional appropriators have used backdoor tactics to reduce that amount.

Environmental Quality Incentives Program (EQIP—1996). Cost share and incentive payments to install or implement structural and management practices. From 1996 to 2002, EQIP was prohibited by law from subsidizing large-scale, regulated industrial livestock confinement operations, and payments were capped at $10,000 per farm per year. Since 2003, EQIP has funded fecal waste management on large-scale CAFOs, with a payment limitation of $450,000 per operation.

Farm and Ranch Land Protection Program (FRPP—1996). Matching funds (up to 50 percent of the fair market value of the conservation easement) to help purchase development rights to keep productive farm and ranchland in agricultural uses. USDA partners with state, tribal, or local governments and nongovernmental organizations to acquire conservation easements or other interests in land from landowners.

Grassland Reserve Program (GRP—2002). Helps landowners restore and protect grassland, rangeland, pastureland, shrubland, and certain other lands. Provides short- or long-term assistance for rehabilitating grasslands, targeting in part vulnerable grasslands from conversion to cropland or development. Grasslands make up the largest land cover on America's private lands—over 525 million acres. The GRP has only enough funding for about 2 million acres.

Wildlife Habitat Incentives Program (WHIP—1996). Provides 5- to 10-year cost-share payments and technical assistance to landowners to develop and improve fish and wildlife habitat primarily on private land. Not limited to agricultural lands. WHIP can reach landowners who are not eligible for other farm conservation programs.

Agricultural Management Assistance (AMA—2000). Cost-share assistance for agricultural producers in 15 states (the northeastern states plus UT, NV, and WY) to construct or improve water management structures or irrigation structures; plant trees for windbreaks or to improve water quality; and mitigate risk through production diversification or resource conservation practices, including soil erosion control, crop rotation, integrated pest management, or transition to organic farming.

Conservation Innovation Grants (CIG—2002). A subset of EQIP. Under CIG, EQIP funds are used to award competitive grants to nonfederal governmental or nongovernmental organizations, Tribes, or individuals to accelerate technology transfer and adoption of promising technologies and approaches to pressing natural resource concerns.

Partnerships and Cooperation Initiative (2002). Allows USDA to designate special projects and enter into stewardship agreements with nonfederal entities, including state and local agencies and nongovernmental organizations, to provide enhanced technical and financial assistance through the integrated application of all the Farm Bill conservation programs. The partnerships help to organize landowners in particular watersheds or defined constituencies and energize them through special incentives to create flexible and efficient solutions to complex resource conservation challenges.

Conservation Technical Assistance (CTA). The CTA Program provides the technical capability, including direct conservation planning, design, and implementation assistance, that helps people plan and apply conservation on the land. A key part of the basic conservation infrastructure for all conservation programs.

Conservation compliance Refers broadly to disincentive programs (Sod Buster, conservation compliance on highly erodible land, and Swamp Buster) created by the 1985 Farm Bill to decrease destructive practices on highly erodible cropland without adequate erosion protection and to prevent the draining of wetlands on agricultural land. Violation of these provisions can result in denial of commodity subsidies and conservation payments, though enforcement has been lax and succeeding farm bills have weakened the rules. Swamp Buster rules are key to achieving the national wetlands no-net-loss policy. Proposals to strengthen Sod Buster are getting renewed attention as limited remaining native prairie gets converted to cropland.

Conservation Security Program

Rewarding the Best, Motivating the Rest

Brian DeVore

On a hot July day Greg Koether leads a group of farmers interested in grass-based livestock production on a tour of one of the hilltop pastures he, his wife Kathy, and their three children farm in northeast Iowa. Koether explains how the field used to be planted to corn and soybeans, and years ago was laced with eroded gullies.

"Now you can't find a gully on the place—I'm pretty proud of that," says Greg as a herd of cattle graze nearby.

In 1983, the Koethers stopped raising row crops and converted the farm to grass, utilizing managed rotational grazing to produce cattle and sheep efficiently while spreading manure in an ecologically sound manner. No tillage has taken place on the Koether farm—they own 500 acres and rent another 238—in over two decades. That was the wrong thing to do, as far as the government was concerned. As the Koethers reduced their plantings of row crops like corn, their federal commodity payments dwindled to around $1,200 a year, a fraction of what a comparably sized cash grain operation receives in the area. But the family persevered, convinced farming methods that put more perennial vegetation on the ground were best for their land, no matter what signals the government sent.

To illustrate just how fragile the landscape is around here, Greg and his teenage daughter Kayla, along with 20-something sons Scott and Klint, take the field day entourage down a road that runs in back of the farm. A working gravel pit dug into the side of a hill provides an impromptu soil profile: a thin layer of topsoil is perched on top of a wall of limestone. The rock is honeycombed with cracks and holes, a textbook example of "karst" geology.

"The limestone rock, despite what some people think, is just a sieve," says Greg. "There's no filtration. There's no water retention. Once that water soaks through that little bit of topsoil, anything in that water is in the water supply." To make things worse, the Koether farm is at the top of the Sny Magill watershed, a coldwater trout stream that is vulnerable to contamination.

By rotating their grazing pastures the Koethers have been able to heal their gullies, stop erosion, and reclaim the land's ability to filter water and keep it from becoming a runoff problem. "We feel the location of our ground in this situation and the production methods we are using is a huge positive impact on the environment," says Greg.

The federal government agrees—finally. After years of being economically punished for not planting corn and soybeans, in 2006 the Koethers qualified for the Conservation Security Program (CSP). Their farm was placed in Tier Three of CSP, the level reserved for the most environmentally friendly farming operations. The fact that their livestock production system relies on year-round grass cover was a major reason the Koethers were able to enroll in CSP. What also helped was that over the years they've established wildlife plantings, windbreaks, and ponds. CSP is also rewarding the family for protecting forestland on the farm and setting up their livestock watering system in a way that reduces animal impact on the land. Some of their stewardship methods, such as rotational grazing, pay off for the Koethers financially. But other things they do to protect the land, such as establishing wildlife habitat, are harder to justify financially. CSP can help bridge the gap between the economic and ecological bottom line.

Through CSP, the Koethers are now receiving around $25,000 annually. "That's quite a change," says Greg. "It's really nice to be rewarded for something that we felt was right all along."

The money will come in handy as the Koethers continue to experiment with methods to make their grazing system even more efficient and make additions to their plantings of wildlife habitat. Kayla Koether says that getting rewarded financially through CSP will make it easier for her and her brothers to eventually take over the operation and farm in a way that's good for the family and the watershed. Seeing operations like the Koether enterprise finally receive a pat on the back from the government sends an important message to young people who may otherwise think the only way to farm is through large-scale row-cropping.

That benefit is not lost on Kayla's father. Says Greg as he wraps up the field day, "If things continue on the right path it's great to realize I'm the third generation here and there could be the fourth and fifth and they could do it all while building the farm, building the soil, and not degrading the environment."

Brian DeVore is the editor of the *Land Stewardship Project Letter* and was a contributor to *The Farm as Natural Habitat: Reconnecting Food Systems with Ecosystems.*

TURNING

THE TABLES

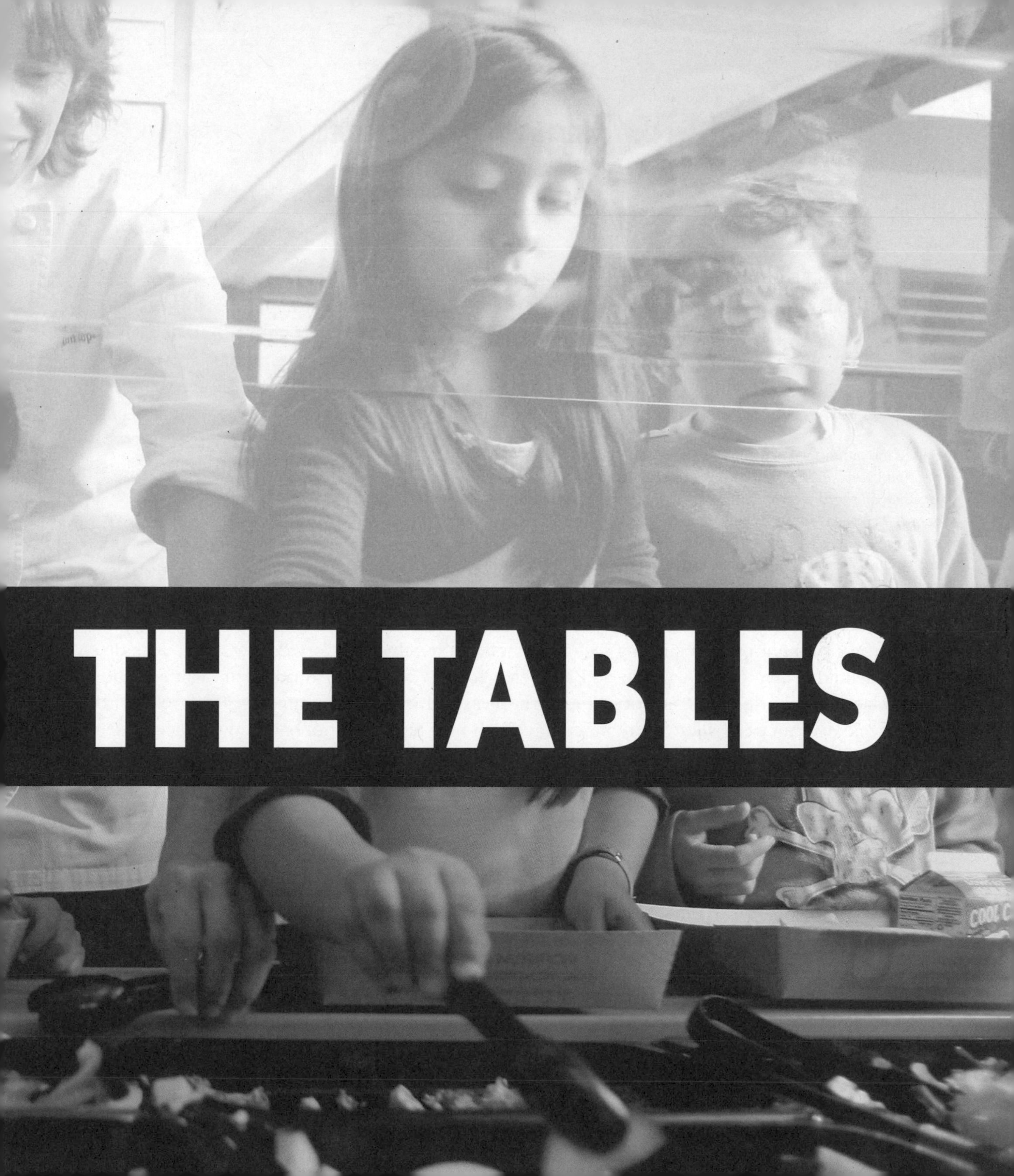

18. Turning the Tables

Food and farm policy is an ongoing cultural and political process, an endless series of give and take from checkout stand to voting booth. But the Farm Bill largely establishes the rules of the game, influencing not only what we eat–but also who grows it, under what conditions, and how much it costs. The agribusinesses, lobbying organizations, and legislators that have essentially written those rules in recent decades deserve the lion's share of the responsibility for shaping the present course of our agriculture and food system. This includes a tangle of critical problems that we have no choice but to address through present and future legislation.

Many of the models we need to turn the tables and avert a collision course for the food and farming system already exist. They all share a common condition: most are either ignored, marginalized, or largely underfunded by Farm Bill programs, yet they surface, a testament to their resilience, tenacity, and ultimately, their solution-oriented wisdom.

Yet people from all walks of life also have enormous influence to bear–as citizens, food consumers, business owners, professionals, doctors, nurses, students, teachers, parents, and community members. Every day, we can support or choose not to support a particular aspect of the food and farming sector through our purchases. Every day, we can speak up for linking family farm health and land stewardship with basic nutritional health in the places we work, in our schools, in our homes, and in our communities. Every election cycle, we can cast ballots for representatives who we hope will not barter away their votes on the Farm Bill or on annual budget appropriations bills, but rather will stand up to preserve what is good in nature and worth preserving or worth changing in our culture.

We can take active leadership positions. Early efforts may seem minimally effective or even symbolic, but later emerge as models to replicate and reinterpret from place to place. Some may even inspire mainstream movements with the ability to redirect food policy at state, national, or even international levels.

The good news is that, to a large extent, many of the ideas we need to turn the tables and avert a collision course for the food and farming system already exist. They all share a common condition: most are ignored, marginalized, or largely underfunded by current Farm Bill programs. Yet they surface, a testament to their resilience, tenacity, and ultimately, their solution-oriented wisdom.

Health Care's Economic Engine

Models such as the on-site farmers' market and CSA distribution programs started by Dr. Preston Maring at Kaiser Permanente Hospital in Oakland, California, are strengthening the bonds between local farmers, organic agriculture, and the employees, patients, and food services of an urban hospital. After years of passing vendors hawking jewelry and leather goods in hospital hallways, Maring–a physician for more than 30 years–convinced management to set up a weekly farmers' market on Kaiser Permanente grounds. He simply wanted to make it easier for people to make healthy food choices by connecting employees and patients with farmers who sell locally grown produce. The effort was an almost instant success. As of late 2006, more than 30 farmers' markets had been established at Kaiser Permanente hospitals in five states (Hawaii, Georgia, Colorado, California, and Oregon), with more on the way. In addition to the farmers' market, a Community Supported Agriculture (CSA) program at the Oakland hospital provides hundreds of employees with weekly deliveries of just-picked fruits and vegetables from a particular farm's harvest–it's a mutually beneficial arrangement for both farmers and employees.

In Madison, Wisconsin, Physicians Plus Insurance Company rewards HMO members who participate in regional Community Supported Agriculture programs with discounts. The "Eat Healthy Rebate Program" deducted $200 for families or $100 for individuals belonging to a CSA in 2006. Physicians Plus' motivation is that prevention (in this case, eating in-season, organic fruits and vegetables) is ultimately the cheapest and most direct path toward good health.[149]

With the largest metropolitan food purchasing population in the country and a potential for significant regional food production in surrounding rural areas, community food organizers are working to bring New York City to the forefront of a new healthy foods movement. FoodChange, a nonprofit antihunger and nutrition advocacy group is among numerous New York City food organizers working to connect urban consumers with the state's family farm community. FoodChange's Harlem SOUL (Sustainable Organic Uptown Local) Food program links rural upstate farmers with low-income inner city residents. Throughout the year, their food pantry and free meal programs feature produce and ingredients sourced directly from small, organic growers. The initiative unites two communities under threat. Statistics show that Harlem residents experience rates of diet-related ailments such as obesity, diabetes, and hyper-tension well above national averages. The New York small farm community is also struggling mightily against rising costs and a centralized global agribusiness industry that makes it very difficult to get their produce to market.

These are just a few examples of models that could be adopted across the country by health professionals and nutrition advocacy organizations, multibillion dollar food purchasing engines with the economic clout to address such interconnected issues as small family farm preservation, health and nutrition, and fossil fuel reduction in one fell swoop. Hard data positively supports these efforts. In the case of Kaiser Permanente, interviews show a nearly 70 percent increase in fruit and vegetable consumption since the farmers' market began.[150] CSA members have been reported to dramatically increase vegetable consumption, by as much as 80 percent among adults, and 60 percent among children.[151]

The Twenty-First Century School Lunch

Ten years ago, renowned chef and Berkeley, California, restauranteur Alice Waters helped to launch a program called the "Edible Schoolyard." Alarmed by both the quantity and quality of outsourced fast foods in the Martin Luther King Middle School, Waters took direct community action. She raised awareness and raised start-up funds. Eventually, a section of the school's playground blacktop was jackhammered away to make room for garden beds. Gardening and more definitively, school lunch, were added to the middle school curriculum. Cafeteria-prepared meals that occasionally featured student-grown produce replaced the mobile fast food wagons and Coke machines. The Edible Schoolyard became a catalyst for a school gardening movement that has swept school districts throughout the state, and increasingly, the country. In 2005, through a grant from the nonprofit Chez Panisse Foundation, the Berkeley Unified School District hired chef Ann Cooper as director of nutrition services with the charge of overhauling the food system for 11 schools, 16 food programs, and 10,000 children.[152] (See Figure 33, *Salad Bars to the Rescue.*) Later, the Berkeley City Council voted to invest in its younger generation and regional farmers by significantly expanding its budget for the purchase of organic foods in all of the city's public schools. In just a few years, Cooper has turned a program nearly dependent on processed ingredients into one where cafeterias cook 95 percent of the meals from scratch.[153]

By declaring the public schools a ground zero for the nutrition, obesity, and family farm crises, Waters and many others around the country committed to school food reform—from Hope, Arkansas, to Santa Monica, California, to Shawnee, Oklahoma, to Atlanta, New Orleans, Philadelphia, and Harlem to name just a few—and have started a transformative movement. A serious, healthy lunch curriculum could ultimately revolutionize education and fitness on par with the President's Physical Fitness programs of the 1960s and 1970s.

Time for Organic's Fair Share

Nearly thirty years ago, organic farmers set out to challenge the application of industrial logic to agriculture and livestock husbandry. Today organic farmers are proving that not only can all crops be grown without harmful chemical inputs and synthetic fertilizers, but their yields can also compete with and in some cases out-produce conventional farming systems, and be healthier for the land in the long-term.[154] Now a rapidly expanding international movement, organic agriculture has experienced the fastest growth of any segment of the food industry for the past ten years, expanding by 20 percent annually, and accounting for 2.5 percent of the market and nearly 25 billion dollars in U.S. sales by 2006. Demand for some organically certified commodities currently outstrips supply in many European countries, where popular support has clearly shifted away from genetically engineered (GMO) crops and toward sustainable agriculture. A number of governments have even established bold targets to dramatically boost the acreage of organic farming across their respective countrysides.

Far from the ominous predictions that a wholesale switch to nonchemical farming methods would result in mass starvation, regionally-based organic agriculture seems like a sound path around which to orient food systems (and therefore Farm Bill policies) of the future. Independent research demonstrates many upsides of organic farming:

- Organic farms have relatively similar or even greater yields than conventional systems depending on place and scale.[155]
- Organic systems use 30 to 70 percent less energy per unit of land than conventional systems, a critical factor in terms of global warming and eventual fossil fuel shortages.[156]
- Organic farms generally support far greater levels of wildlife, particularly in comparison with large-scale intensive agriculture.[157]
- Organic foods distributed through local and regional distribution chains offer reduced energy consumption, less processing and packaging, and higher nutritional values.
- Organic farmers selling to local markets are more likely to grow rare breeds and varieties carefully selected for their specific growing conditions (rather than shipability, yield, and uniformity).

Despite these benefits, just 0.35 percent of the $1 billion Agricultural Research Service budget is directed to organic farming issues. In terms of market share, this is a proportional underfunding of 700 percent, according to Bob Scowcroft, director of the Organic Farming and Research Foundation (OFRF) in Santa Cruz, California. Agencies such as the Economic Research Service, Ag Outlook, and Risk Management have also been slow to respond to the growth and importance of the industry.

Loyal organic consumers—there are literally millions of them—have plenty of room to be disgruntled with this taxation without equitable representation. As taxpayers, they contribute to subsidies that largely go to conventional farming. At the cash register, they often pay more for organic foods, as their farm constituencies are shut out of the price support game. Finally, they're taxed a third time when government spending is required to address problems related to conventional industrial farming (such as water cleanup, health problems from pesticides and agrichemical pollution, and so on.) Look for organic advocates to increasingly demand and defend their fair share in all aspects of food and farm policy.

Grass Land—A Farm Bill Imperative

Problems related to feedlot agriculture have already been outlined in-depth: manure containment crises, antibiotics and growth hormones passed on to humans through meat and dairy products, environmental degradation due to corn and soybean monoculture farming, the spread of infectious and even lethal diseases to both animals and humans. The list goes on.

The elegant solution to grainfed, confinement-raised livestock is simple: turn the cattle out of the animal gulags and make it possible for them to eat grass. Of course that will drastically reduse the need for the decades-old taxpayer-subsidized below-cost corn and soybean feed supply, which for so long has kept small producers at a disadvantage. Yet long before the row-crop revolution, grasslands dominated much of the United States, serving as a matrix for most of the continent's natural ecological processes—precipitation, pollination, infiltration, predation, migration, respiration, decomposition, and mineralization.[158] Herbivores evolved to eat grass, which also protected the soil, survived droughts, and naturalized to local conditions and environments. Before the advent of the mega-farm, diversified family farms included animals and pastures, a practical arrangement that provided a ready source of fertilizer (manure) for plants as well as meat and dairy products for local and regional markets.

Transitioning toward grass farming and perennial agriculture—deep-rooted crops that do not need to be replanted every year—should be seriously regarded as a fundamental foundation for a new era of agriculture and food and farm policy. Pushed to the brink by subsidized sod busting, grassland birds such as vesper sparrows, bobolinks, meadowlarks, and other species may disappear from traditional ranges without the restoration of grasslands to Midwestern farm fields. From a human health perspective, it has been widely documented that animals allowed to graze produce healthier dairy products and leaner meats, higher in beneficial Omega-3 fatty acids than their grainfed factory grown counterparts. Grass farming makes sound economic sense as well. With its limited start-up expenses (land, shelter, and portable fencing), it is becoming a relatively inexpensive route for young farmers ambitious enough

Figure 33

Salad Bars to the Rescue

Daily servings of fruits and vegetables

4.5
4
3.5
3
2.5
2
1.5
1
0.5
0

2000 post-Salad Bar

1998 pre-Salad Bar

Source: "Evaluation of the Effectiveness of the Salad Bar Program in the Los Angeles Unified School District."

Proof is in the Produce. Offering farm-fresh, good tasting produce makes kids want to eat it. At least that's what the Los Angeles Unified School District found out through a review of students' eating behavior before and after the installation of an in-school salad bar. In just three months, students increased the amount of fruits and vegetables they ate by an average of 1¼ servings per day. They also decreased their lunchtime fat intake by 200 grams.

The Berkeley Unified School District has witnessed a far-reaching transformation of its food service in the past few years. Ann Cooper, in the photo above, oversees 11 school gardening programs and 16 food programs, impacting 10,000 children. Here are just a few features of Berkeley Unified's school lunch revolution*:

- 95 percent of meals are made from scratch.
- Reduced-meal students pay with a swipe-card system.
- Only hormone-free and antibiotic-free dairy is served.
- Fresh fruits and vegetables are served at every meal.
- Most food is sourced from locally-owned businesses.
- These changes add 50 cents per lunch (food and labor costs) above federal and state reimbursements.

*Anna Lappé, "Doing Lunch: Ann Cooper Serves up a New Vision of School Food," *The Nation*, September 11, 2006, p. 35.

to enter agriculture as a profession. Even so, the infrastructure necessary to support such a shift–regional and mobile slaughtering facilities, processing centers, and the like–have largely disappeared from the rural areas and must now be reestablished. Many now hope that grass farming and other perennial systems could lead to far more diversified forms of agriculture, producing a greater variety of grains, oils, and fruits, as well as meats and dairy, and far fewer sweeteners and unhealthy hydrogenated oils.

Climate Change: The Ultimate Wild Card

Scientists expect that changes in the global climate will severely impact agricultural systems around the world, if not in the short term, then some time later this century. Unpredictable and violent storm events are projected, resulting in flooding, soil loss, and other potentially catastrophic conditions. Researchers have predicted that southern and plains states may see crop reductions with lower precipitation and higher temperatures. Other regions of the United States will experience changes in the frequency and intensity of droughts, variable rainfall patterns, glacial melting, and alterations in plant and animal communities. If only to build more resilience within farming landscapes, the shift away from intensive row-cropping systems (which contribute the most greenhouse gases and are most vulnerable to flooding and erosion) to perennial systems (the most effective carbon capturers and soil stabilizers) deserves nationwide attention along with an aggressive policy commitment.

Healthy Lands, Healthy Tables

Thirty years ago, few of us could have predicted the accelerated rise in demand of organic foods today. Ten years ago, Japanese car manufacturers might have been considered delusional for taking the costly leap into the development of hybrid vehicle technology. Yet these innovations, in response to shifting economic, environmental, and cultural conditions, have proven both prescient and economically insightful.

Conditions now call for an equally bold new direction for food and farm policy. Common sense demands that narrow self-interested program development yield to an updated and broader vision. Local and regional production and distribution capabilities should be greatly up-scaled and expanded immediately, if not to cut down on food miles, then to preserve family farms; if not to curb global warming by reducing energy use, then to provide more nutritious foods to local and regional markets; if not to encourage more geographic equity among subsidy recipients, then to strengthen both national security and local food security; if not to prevent sprawl and the loss of open space, then to invest in the potential for rural areas as tourist destinations.

Just as the Farm Bill should evolve to include Food in its name and its mission, the

Farm Bill's program titles must become more deeply integrated. All government subsidies should pivot around the essential goal of conservation–whether commodity agriculture, energy, research, or any other activity.

Maybe it's time to question whether the faraway industrial mega-farm model is indeed inevitable, or preferable, or even sustainable without costly government supports. Perhaps it's time that citizens begin to see that Farm Bill politics–as the saying goes–are local politics. Perhaps a farm and food policy that is taking a toll on the land, making the population overweight and obese, and tearing away at the fabric of rural communities, requires an era of new solutions.

Perhaps the time has arrived for a food fight.

25 Ideas Whose Time Has Come

1. More closely align Farm Bill crop supports with USDA nutritional guidelines.
2. Fully fund, expand, and refine the Conservation Security Program that rewards stewardship and sound farming rather than surplus production.
3. Design subsidies to function more as safety nets, loans, crop reserves, and stewardship incentives rather than direct giveaways.
4. Establish an effective cap limit on individual subsidy recipients and close loopholes to provide greater equity to all farmers.
5. Shorten the food mile component of the current farming and distribution system by rebuilding the infrastructure for community-based and regional food supply chains.
6. Expand affordability and access to high-quality healthy foods for everyone, particularly in rural and urban communities where resources and options are limited.
7. Launch a national healthy lunch and fitness program that generates incentives for local and regional farms, features a salad bar and school gardening program in every school, and strives for meals made from scratch.
8. Keep small farmers on the land by working for fair prices for all crops.
9. Expand farm and ranchland preservation programs that buffer communities against sprawl, maintain habitats, and keep valuable agricultural lands in production.
10. Provide expanded funding for the preservation (as well as strict penalties for the plowing) of remnant native prairies and functional grasslands.
11. Shift incentives away from corn- and soybean-based feedlots and toward a grass-based livestock economy.
12. Target research and incentives toward reductions in farm-related global warming emissions, such as organic and perennial agriculture.

13. Include global warming reduction goals in Farm Bill titles whenever possible.
14. Restore incentives, start-up loans, and respectability for future generations of farmers and food producers through beginning farmer programs.
15. Establish conservation standards for Farm Bill-funded ethanol or other bio-based energy crops.
16. Expand set-aside programs (Conservation Reserve Program, Wetland Reserve Program, Grassland Reserve Program) with nationwide goals for restoration and watershed protection.
17. Fund more on-the-ground technical conservation assistance and enforcement.
18. Increase oversight and accountability of taxpayer funded crop insurance programs.
19. Promote growth in farmers' markets throughout the country as well as farm-to-school, farm-to-hospital, farm-to-health care provider, and other farm-direct distribution arrangements.
20. Increase funding for farm-scale and utility-scale renewable energy projects, with continued emphasis on conservation and energy efficiency.
21. Expand research into energy-saving agriculture methods, including alternatives to synthetic fertilizers.
22. Continue to work toward campaign finance and lobbying reform.
23. More closely integrate and reward forest owners as part of the agricultural landscape.
24. Include native pollinator habitat restoration and invasive species removal as regional and nationwide conservation goals.
25. Integrate food and farm policy goals with other key legislative programs: transportation, energy, health, national defense, immigration, minimum wages, and so on.

Join the Food Fight

Flex Your Vote–Write your representatives and insist that they not trade their votes on Farm Bills or annual spending bills related to food and farm policy.

Get Up, Stand Up–Continue to press for campaign-finance and lobbying reform as well as anti-trust enforcement to challenge Big Food's influence on food and farm policy.

Build Bridges–Try to understand the concerns of other groups in your area with respect to the Farm Bill. Look for unifying public benefits such as renewable energy, a vibrant family farm economy, hunger relief, school nutrition, the protection of open space, and so on.

Make Schools Healthy–Get involved in nutrition and gardening programs in your local schools. Local efforts can influence a district, which reaches a city and starts a chain reaction that could ultimately lead to federal policy changes.

Make Food Miles Count–Support local farmers and community food chains whenever possible–through farmer's markets, Community Supported Agriculture, and other farmer direct distribution programs.

Insist on Grass–Purchase pasture-raised dairy and livestock products when possible. Insist that your representative have a policy position on grass farming and grassland restoration.

Fight Hunger with Healthy Food–Collaborate with your local community to bring healthy food to those who need it the most through youth outreach programs, elder hospices, farmer's markets, farm-direct distribution programs, and other creative avenues.

Keep Congress Accountable–After a Farm Bill is passed into law, keep legislators honest by making sure annual budgets actually support the programs that have been promised.

Be a Proud Conservationist–Long-term land health must become the basis of all farm subsidy programs so that future generations inherit sound environments.

Nutrition Not Just Calories–Insist that farm subsidies support a diversified and nutritious diet rather than just cheap feeds and ingredients for industrially processed foods.

Take the Lead–Use your influence at work, at home, and in the community to create a change in the food and farming system for the better.

Don't Act Alone–Join the Farm Bill food fight in your area by participating with groups working on long-term plans for food security and food autonomy in your local community and region. Connect with others in your region to protect good pieces of legislation and prevent egregious bills from passing.

The Midwest Sustainable Agriculture Working Group website (www.msawg.org.) provides an updated list of groups to get involved with: if you live in Illinois, consider joining forces with the Illinois Stewardship Alliance; in Minnesota, the Land Stewardship Project; in Kansas, the Kansas Rural Center; and so on. Other groups include:

- Community Food Security Coalition (www.foodsecurity.org)
- Environmental Working Group (www.ewg.org)
- Institute for Agriculture and Trade Policy (www.iatp.org)
- Land Stewardship Project (www.landstewardshipproject.org)
- National Campaign for Sustainable Agriculture (www.sustainableagricuture.net)
- National Catholic Rural Life Conference (www.ncrlc.com)
- The Rural Advancement Foundation International (www.rafiusa.org)
- Western Organization of Resource Councils (www.worc.org)
- California Coalition for Food and Farming (www.calfoodandfarming.org)

Resources

Assembled here is a list of organizations, agencies, educational institutes, and other entities to help you find models, make connections, research policy platforms, and learn much more about Food and Farm Bill issues. Excellent reporting in the Washington Post, Atlanta Journal-Constitution, Wall Street Journal, New York Times, and other avenues can also provide updated analysis on a wide variety of issues. Be sure to follow the Watershed Media website (www.watershedmedia.org) for updated listings of valuable resources.

Conservation Groups

Defenders of Wildlife
Washington, D.C.
(800) 385-9712
www.defenders.org
Dedicated to the protection of all native wild animals and plants in their natural communities.

Environmental Defense
New York, NY
(212) 505-2100
www.
environmentaldefense.org
Dedicated to protecting the environmental rights of all people, now and in the future. Among these rights are clean air, clean water, healthy food and flourishing ecosystems.

Natural Resources Defense Council
New York, NY
(212) 727-2700
www.nrdc.org
Dedicated to protecting the long-term welfare of present and future generations of all life on earth while working to restore air, land and water.

Soil and Water Conservation Society
Ankeny, IO
(515) 289-2331
www.swcs.org
A nonprofit scientific and educational organization that serves as an advocate for science-based conservation practices, programs, and policy.

Wild Farm Alliance
Watsonville, CA
(831) 761-8408
www.wildfarmalliance.org
Nonprofit outreach and research organization dedicated to promoting healthy agriculture operations and conservation efforts that protect wild nature.

Xerces Society
Portland, OR
(503) 232-6639
www.xerces.org
Advocates for invertebrates and their habitats by working with scientists, land managers, educators, and citizens on conservation and education projects.

Sustainable Agriculture

American Farmland Trust
Washington, D.C.
(202) 331-7300
www.farmland.org
A national nonprofit organization working with communities and individuals to protect the best land, plan for growth, and keep the land healthy.

Center for Rural Affairs
Lyons, NE
(402) 687-2100
www.cfra.org
A private nonprofit that is working to strengthen small businesses, family farms, ranches, and rural communities.

Community Alliance with Family Farmers
Davis, CA
(530) 756-8518
www.caff.org
Building a movement of rural and urban people to foster family-scale agriculture that cares for the land, sustains local economies, and promotes social justice.

Ecological Farming Association
Watsonville, CA
(831) 763-2111
www.eco-farm.org
Brings together growers, consumers, educators, activists, and industry related businesses to exchange the latest advances in sustainable food production and marketing.

Glynwood Center
Cold Spring, NY
(845) 265-3338
www.glynwood.org
Creates programs to train leaders, catalyze community change, and supply resources to those who strive to resolve the tension between development and conservation.

Land Stewardship Project
White Bear Lake, MN
(651) 653-0618
www.
landstewardshipproject.org
A grassroots organization of farmers, rural, and urban residents working together to foster an ethic of stewardship for farms, sustainable agriculture, and sustainable communities.

ily farmers, safe food, and vibrant rural communities.

socially just, and produces safe, nutritious food.

tion before Congress and federal administrative agencies.

Michael Fields Agricultural Institute
Troy, WI
(262) 642-3303
www.michaelfieldsaginst.org
Cultivates the ecological, social, economic, and spiritual vitality of food and farming systems through education, research, policy, and market development.

Midwest Sustainable Agriculture Working Group
Indianapolis, IN
(317) 536-2315
www.msawg.org
A network of organizations working for a system of agriculture that is economically profitable, environmentally sound, family-farm based, and socially just.

National Campaign for Sustainable Agriculture
Pine Bush, NY
(845) 361-5201
www.sustainableagriculture.net
Dedicated to educating the public on the importance of a sustainable food and agriculture system that is economically viable, environmentally sound, socially just, and humane.

National Family Farm Coalition
Washington, D.C.
(202) 543-5675
www.nffc.net
A voice for grassroots groups on farm, food, trade, and rural economic issues to ensure fair prices for fam-

National Information System for Regional IPM Centers
Raleigh, NC
(530) 754-8378
www.wrpmc.ucdavis.edu
This site provides information about commodities, pests and pest management practices in the U.S.

National Sustainable Agriculture Information Service
Fayetteville, AR
(800) 346-9140
www.attra.ncat.org
Provides information and other technical assistance to farmers, ranchers, extension agents, educators, and others involved in sustainable agriculture in the United States.

Northeast-Midwest Institute
Washington, D.C.
(202) 544-5200
www.nemw.org
A private, non-profit, and nonpartisan research organization dedicated to economic vitality, environmental quality, and regional equity for Northeast and Midwest states.

Northeast Sustainable Agriculture Working Group
Belchertown, MA
(413) 323-9878
www.nesawg.org
Works to build a sustainable regional food and agriculture system that is environmentally sound, economically viable,

Organic Trade Association
Greenfield, MA
(413) 774-7511
www.ota.com
A membership-based business association working to encourage global sustainability through promoting and protecting the growth of diverse organic trade.

Pesticide Action Network North America
San Francisco, CA
(415) 981-1771
www.panna.org
PANNA challenges the global proliferation of pesticides, defends basic rights to health and environmental quality, and works to insure the transition to a just and viable society.

Sierra Club National Agriculture Committee
San Francisco, CA
(415) 977-5500
www.sierraclub.org/agriculture/committee
Works with members of the Sierra Club and the public for a more sustainably balanced and equitable system of agriculture.

Sustainable Agriculture Coalition
Washington, D.C.
(202) 547-5754
www.msawg.org
SAC takes a common position on critical agricultural and environmental policy and provides support for representa-

Sustainable Agriculture & Food Systems Funders (SAFSF)
Santa Barbara, CA
(805) 687.0551
www.safsf.org
A national working group of grantmakers that works to promote a more sustainable agriculture and food system.

Community Food Security and Anti-Hunger

America's Second Harvest
Chicago, IL
(800) 771-2303
www.secondharvest.org
A national food bank network that distributes food, increases public awareness of domestic hunger, and advocates public policies that benefit America's hungry.

Bread for the World
Washington, D.C.
(202) 639-9400
www.bread.org
Faith-based advocacy organization that seeks justice for hungry people through research and education on hunger and development.

California Food Policy Advocates
San Francisco, CA
(415) 777-4422
www.cfpa.net
An advocacy organization dedicated to improving the health and well being of low-income

Californians by increasing their access to nutritious and affordable food.

California Coalition for Food and Farming
Watsonville, CA
(831) 763-2111
www.calfoodandfarming.org
An active network dedicated to promoting a sustainable and socially just food system.

Center for Food Safety
Washington, D.C.
(202) 547-9359
www.centerforfoodsafety.org
A non-profit public interest and environmental advocacy organization that challenges harmful food production technologies and promotes sustainable alternatives.

Community Food Security Coalition
Venice, CA
(310) 822-5410
www.foodsecurity.org
An organization representing issues of social and economic justice, environment, sustainable agriculture, community development, anti-poverty, and anti-hunger.

Congressional Hunger Center
Washington, D.C.
(202) 547-7022
www.hungercenter.org
Serves as a center where the anti-hunger community can discuss creative solutions to end domestic and international hunger.

Food Change
New York, NY
(212) 894-8094
www.foodchange.org
Improving lives through innovative nutrition and education programs and financial empowerment.

Food Routes
Arnot, PA
(814) 349-6000
www.foodroutes.org
A national organization that promotes "Buy Fresh, Buy Local" campaigns by providing communications tools, marketing resources, and support to grassroots groups in the U.S.

Food Security Learning Center
New York, NY
(212) 629-8850
www.worldhungeryear.org/fslc/
An extension of the World Hunger Year movement that focuses on community food security, rural poverty, hunger, nutrition, family farms and migrant workers.

Growing Power
Milwaukee, WI
(414) 527-1546
www.growingpower.org
A nationwide nonprofit organization developing community food systems that provide high-quality, safe, healthy, affordable food for all residents in the community.

National Catholic Rural Life Center
Des Moines, IA
(515) 270-2634
www.ncrlc.com
Works for justice, solidarity, and the integrity of creation in order to insure safe food systems, to defend nature, and keep families on the land.

National Council of Churches
New York, NY
(212) 870-2227
www.ncccusa.org
A leading force for cooperation among Christians in the United States, on programs dealing with hunger, land stewardship, environmental protection, poverty, and health.

The Food Trust
Washington, D.C.
(215) 568-0830
www.thefoodtrust.org
Works to improve the health of children and adults, promote good nutrition, increase access to nutritious foods, and advocate for better public policy.

Education

Center for Ecoliteracy
Berkeley, CA
(510) 845-4595
www.ecoliteracy.org
A public foundation that supports a grant-making program for educational organizations and school communities in the San Francisco Bay Area.

Leopold Center for Sustainable Agriculture
Aimes, IO
(515) 294-3711
www.leopold.iastate.edu
Identifies and reduces the negative impacts of agriculture on natural resources and rural communities.

UC Sustainable Agriculture Research and Education Program
Davis, CA
(530) 752-7556
www.sarep.ucdavis.edu
Concerned about the environmental impacts of agriculture, the health of rural communities, and the profitability of family farming operations in California.

USDA Sustainable Agriculture Research and Education
Washington, D.C.
(202) 720-5384
www.sare.org
SARE helps to advance farming systems that are profitable, environmentally sound, and good for communities through a nationwide research and education grants program.

International Trade and Globalization

Forum on Democracy and Trade
(718) 797-9472
www.forumdemocracy.net
U.S. public officials concerned about the ways in which trade

and investment agreements impact local democratic decision-making and economic development.

International Forum on Globalization
San Francisco, CA
(415) 561-7650
www.ifg.org
An alliance of activists, scholars, economists, researchers, and writers formed to stimulate new thinking and public education in response to economic globalization.

International Society for Ecology and Culture
Berkeley, CA
(510) 548-4915
www.isec.org.uk
Works for the protection of cultural and biological diversity with an emphasis on moving beyond single issues to look at the fundamental influences that shape lives.

Oxfam
Oxford, UK
44 (0) 1865 473727
www.oxfam.org.uk
A development, relief, and campaigning organization that works with others to find lasting solutions to poverty and suffering around the world.

World Trade Organization
Geneva, Switzerland
(41-22) 739 5111
www.wto.org
The global international organization that deals with the rules of trade between nations.

Renewable Energy

Energy Foundation
San Francisco, CA
(415) 561-6700
www.ef.org
A partnership of major donors working to advance energy efficiency and renewable energy through new technologies that are essential components of a clean energy future.

Environmental Law & Policy Center
Chicago, IL
(312) 673-6500
www.elpc.org
Works to achieve cleaner energy resources and implement sustainable energy strategies, by promoting innovative and efficient transportation and land use approaches.

The Minnesota Project
St. Paul, MN
(651) 645-6159
www.mnproject.org
Focused on clean renewable energy and efficient energy, farm practice that is profitable and protects the environment, and local, sustainably produced foods.

Government Agencies

Association of State and Interstate Water Pollution Control Administrators
Washington, D.C.
(202) 756-0600
www.asiwpca.org
State and Interstate officials responsible for the implementation of surface water protection programs throughout the nation.

Congress
www.congress.org
See above site for instructions on how to write to your representative.

Congressional Research Service
Washington, D.C.
www.loc.gov/crsinfo/
A branch of the Library of Congress that prepares nonpartisan reports for Congress. CRS reports can be found at these websites:
University of North Texas
www.digital.library.unt.edu/govdocs/crs/
National Council for Science and Environment
www.ncseonline.org/NLE/CRS/

Economic Research Service
Washington, D.C.
(202) 694-5050
www.ers.usda.gov
Source of government oriented reports on agriculture and natural resources.

Natural Resources Conservation Service
Washington, D.C.
See website for contact info
www.nrcs.usda.gov
The conservation arm of the USDA that works to help landowners conserve, maintain, and improve our natural resources and environment.

United States Department of Agriculture
Washington, D.C.
See website for contact info
www.usda.gov

USDA Farm Bill
www.usda.gov/farmbill
Forums and analyses regarding the 2007 Farm Bill.

Policy

California Institute for the Study of Specialty Crops
San Luis Obispo, CA
(805) 756-2855
www.cissc.calpoly.edu
Created to advance research, discourse, and educational opportunities related to agricultural policy.

Chicago Council on Global Affairs
Chicago, IL
(312) 726-3860
www.thechicagocouncil.org
An independent, nonpartisan organization committed to influencing the discourse on global issues through policy formation, leadership dialogue, and public learning.

Environmental Working Group
Washington, D.C.
(202) 667-6982
www.ewg.org
A team of civil servants and professionals who analyze data and expose threats to your health and environment, and seeks solutions to these problems.

Food First

Oakland, CA
(510) 654-4400
www.foodfirst.org

Food First is a think tank and action center that highlights root causes and value-based solutions to hunger and poverty around the world.

Food Research and Action Center

Washington, D.C.
(202) 986-2200
www.frac.org

Works to improve public policy to eradicate hunger and under-nutrition in the US.

Institute for Agriculture and Trade Policy

Minneapolis, MN
(612) 870-0453
www.iatp.org

Promotes resilient family farms, rural communities, and ecosystems around the world through research and education, science and technology, and advocacy.

Organic Farming and Research Foundation

Santa Cruz, CA
(831) 426-6606
www.ofrf.org

OFRF works to support organic farmers' immediate information needs while moving the public and policymakers toward greater investment in organic farming systems.

Public Citizen

Washington, D.C.
(202) 588-1000
www.citizen.org

A consumer advocacy organization that represents consumer issues in Congress, the Executive branch, and the Judicial branch of government.

Rudd Center for Food Policy and Obesity

New Haven, CT
(203) 432-6700
www.yaleruddcenter.org

Strives to improve practices and policies related to nutrition and obesity that inform and empower the public and maximize the impact on public health.

Southern Legislative Conference

Atlanta, GA
(404) 633-1866
www.slcatlanta.org

Works to foster intergovernmental cooperation among member states while addressing issues such as agriculture, rural development, economic development, and transportation.

W.K. Kellogg Foundation

Battle Creek, MI
(269) 968-1611
www.wkkf.org

Funds programs that promote a world in which each person accepts responsibility for self, family, community, and societal well-being, and strives to help create nurturing families, responsive institutions, and healthy communities.

World Resources Institute

Washington, D.C.
(202) 729-7600
www.wri.org

An environmental think tank that focuses on conservation research that will foster environmentally sound, socially equitable development.

Health and Nutrition

California Nutrition Network

www.dhs.ca.gov/ps/cdic/cpns/network

Creates innovative partnerships so that low-income Californians are enabled to adopt healthy eating and physical activity patterns as part of a healthy lifestyle.

Society of Nutrition Educators

Indianapolis, IN
(800) 235-6690
www.sne.org

Advances the food and nutrition education profession to impact healthful food choices and lifestyle behaviors at the individual, community and policy levels.

Select Bibliography

"An American Success Story: The Farm Bill's Clean Energy Programs." Environmental Law and Policy Center.

"Are You Being Served? Environmental Entries Are Starting to Appear on the Balance Sheet." *Economist*, April 22, 2005, pp. 76–78.

Baker, David R. "Got Gas? PG&E Considers Using a Natural Energy Source That Comes from Cows–And It Sure Isn't Milk." *San Francisco Chronicle*, July 12, 2006, p. C1–C5.

Bales, Susan Nall. "Framing the Food System: A Frame Works Message Memo." Paper delivered at the Sixth Annual W. K. Kellogg Foundation Food and Society Networking Meeting, April 24–26, 2006.

Barrinuevo, Alexei. "For Good or Ill, Boom in Ethanol Reshapes Economy of Heartland." *New York Times*, June 25, 2006, web edition.

Barta, Patrick, and Jane Spencer. "Crude Awakening: As Environmental Energy Heats Up Environmental Concerns Grow." *Wall Street Journal*, December 5, 2006.

Bowers, Douglas, Wayne Rasmussen, and Gladys Baker. "History of Agricultural Price-Support and Adjustment Programs, 1933–84." Agricultural Information Bulletin No. AIB485, December 1984.

Burnett, John. "Fraud on the Farm: Tomato Farmers Caught Out in Insurance Scam." National Public Radio Morning Edition, April 12, 2006, transcript.

Carlson, Scott. "For Amber Waves of Grain: The 2007 Farm Bill: Who Is Gearing up for a Food Fight, and Why You Should Care." *Utne Reader*, March–April 2006, pp. 79–83.

Charles, Dan. "Going Green in Agriculture: Calls Grow to Subsidize Green Farming." National Public Radio Morning Edition, April 13, 2006, transcript.

——. "Going Green in Agriculture: EU Shifts Subsidies from Crops to Land Stewardship." National Public Radio Morning Edition, April 13, 2006, transcript.

"Charm of the Farm, The," Editorial. *Washington Post*, October 19, 2005, washingtonpost.com/wp-dyn/content/article/2005/10/18/AR2005101801487.html.

Cleeton, James. "Organic Foods in Relation to Nutrition and Health: Key Facts." Information Sheet summary of "Coronary and Diabetic Care in the UK2004," by the Association of Primary Care Groups and Trusts, Soil Association, United Kingdom.

Cook, Christopher. "Business as Usual." *The American Prospect On-line*, April 8, 2006.

Defenders of Wildlife. "Defenders of Wildlife's Comments for the Development of USDA Recommendations for the 2007 Farm Bill (70 Federal register 35221)," June 17, 2005.

Deutch, John. "Biomass Movement," Commentary. *Wall Street Journal Online*, May 10, 2006, http://online.wsj.com/article_email/SB114722621580248526-lMyQjAxMDE2NDE3MDIxMjA2Wj.html.

Diamond, Jared. *Guns, Germs, and Steel: The Fates of Human Societies*. W.W. Norton, 1999.

Eaton, Sam. "Dead Zone Threatens Shrimp Industry." *Marketplace*, Thursday, July 13, 2006, transcript, http://marketplace.publicradio.org/shows/2006/07/13/PM200607139.html.

Environmental Law and Policy Center. "An American Success Story: The Farm Bill's Clean Energy Programs."

"The Ethanol Myth: Consumer Reports' E85 Tests Show that You'll Get Cleaner Emissions but Poorer Fuel Economy…If you Can Find It." *Consumer Reports*, October 2006, pp. 15–19.

Foskett, Ken, and Dan Chapman. "Farmers Accused of 'Scheme' Wind up Keeping Millions." *Atlanta Journal-Constitution*, Oct. 3, 2006.

Foskett, Ken, Dan Chapman, and Megan Clarke. "How Savvy Growers Can Double, or Triple, Subsidy Dollars." *Atlanta Journal*-Constitution, Oct. 2, 2006.

——. "How Your Tax Dollars Prop Up Big Growers and Squeeze the Little Guy." *Atlanta Journal-Constitution*, Oct. 1, 2006.

Galbraith, John Kenneth. "The Predator State: Enron, Tyco, World Com…and the U.S. Government?" *Mother Jones*, May June 2006, pp. 30–31.

Greene, Nathanael. "Growing Energy: How Biofuels Can Help America's Oil Dependence." Natural Resources Defense Council, December 2004.

Heilprin, John. "U.S. Reports Increase in Wetland Acreage: Bush Administration Figures Are Disputed as Being Misleading." *San Francisco Chronicle*, March 31, 2006, p. A2.

Huber, Peter W., and Mark P. Mills. *The Bottomless Well: The Twilight of Fuel, the Virtue of Waste, and Why We Will Never Run Out of Energy*. New York: Basic Books, 2005.

Hull, Jonathan Watts. "The 2007 Farm Bill in Context: Summer 2006 Update." Southern Legislative Conference, June 2006, web edition.

Kasler, Dale. "Ethanol Boom Lures Investors: Some Put Their Money into In-State Production Plants." *Sacramento Bee*, March 22, 2006, A1 and A20.

Kilman, Scott, and Roger Throw. "In Fight Against Farm Subsidies, Even Farmers Are Joining Foes." *Wall Street Journal*, March 14, 2006, www.truthabouttrade.org/article.asp?id=5398.

Kinney, Donal. "Ethanol: A Transitional Strategy Toward Sustainability." *Coop Connection* (Manzanita Coop Newsletter, Albuquerque, New Mexico), April 2006, pp. 2 and 14.

Kotschi, Johannes, and Karl Müller-Sämann. "The Role of Organic Agriculture in Mitigating Climate Change: A Scoping Study." International Federation of Organic Agriculture Movements, May 2004.

Kunstler, James Howard. *The Long Emergency: Surviving the Converging Catastrophes of the Twenty-first Century*. New York: Atlantic Monthly Press, 2005.

Leahy, Stephen. "Population: Global Food Supply Near the Breaking Point." May 17, 2006, Inter Press Service News Agency, http://www.ipsnews.net/news.asp?idnews=33268.

Lilley, Ray. "A Real Job: End to Subsidies Improved Farming in New Zealand." December 20, 2005, The Associated Press, www.journalnow.com.

Lynch, Sarah, and Sandra Batie, eds. "Building the Scientific Basis for Green Payments." A Report on a Workshop Sponsored by the U.S. Department of Agriculture, CSREES, World Wildlife Fund, and the Elton R. Smith Endowment at Michigan State University, April 14–15, 2005, World Wildlife Fund, Washington D.C.

Manning, Richard. *Against the Grain: How Agriculture Highjacked Civilization*. North Point Press, 2004.

——. "The Oil We Eat: Following the Food Chain Back to Iraq." *Harper's* Magazine, February 2004.

Marlow, Scott. "The Non-Wonk Guide to Understanding Federal Commodity Payments." The Rural Advancement Foundation International, 2005.

Marshall, Liz, and Suzie Greenhalgh. "Beyond the RFS: The Environmental and Economic Impacts of Increased Grain Ethanol Production in the U.S." WRI Policy Note, World Resources Institute, Washington D.C., September 2006.

Millman, Joe. "Labor Movement: As U.S. Debates Guest Workers, They Are Here Now in Construction 'Subidos.'" *The Wall Street Journal*, September 18, 2006, A1.

Morgan, Dan, and Gilbert M. Gaul. "No Farm? No Crop? No Problem–Subsidies Still Grow." *Washington Post*, July 2, 2006, web edition.

Muller, Mark, and Heather Schoonover. "U.S. Farm Policy Contributes to Obesity, New Report Finds Policies Drive Production of Unhealthy Foods." Institute for Agriculture and Trade Policy, 2006.

Nestle, Marion. *Food Politics: How the Food Industry Influences Nutrition and Health.* University of California Press, 2002.

"No Time for Delay: A Sustainable Agriculture Agenda for the 2007 Farm Bill." Sustainable Agriculture Coalition, October 2006.

O'Brien, Doug, Halley Torres Aldeen, Stephanie Uchima, and Errin Staley. "Hunger in America: The Definitions, Scope, Causes, History and Status of the Problem of Hunger in the United States." America's Second Harvest, Public Policy and Research Department, 2004.

"Out of Balance: Marketing of Soda, Candy, Snacks, and Fast Foods Drowns Out Healthful Messages." Consumers Union and California Pan-Ethnic Health Network, September 2005

"Panel Debates Next Farm Bill's Impact on California." *California Agriculture* 60, no. 1 (January–March 2006): 5–7, CaliforniaAgriculture.uocp.edu

Planck, Nina. "Leafy Green Sewage." *The New York Times*, September 21, 2006.

Pollack, Andrew. "Redesigning Crops to Harvest Fuel: Scientists as Custom Tailors of Genetics." *New York Times*, September 8, 2006, C1–C4.

Pollan, Michael, *The Omnivore's Dilemma: A Natural History of Four Meals.* Penguin, 2006.

——. "Six Rules for Eating Wisely." *Time* Magazine, June 12, 2006, p. 9.

Pretty, Jules. "The Real Costs of Modern Farming: Pollution of Water, Erosion of Soil and Loss of Natural Habitat, Caused by Chemical Agriculture, Cost the Earth." *Resurgence*, issue 205, web edition.

Rooney, William, ed. *Fish and Wildlife Benefits of Farm Bill Conservation Programs, 2000–2005 Update.* USDA Natural Resources Conservation Service and Farm Service Agency.

Roth, Dennis. "Food Stamps: 1932–1977: From Provisional and Pilot Programs to Permanent Policy." Economic Research Service, web edition.

Shepherd, Matthew, Stephen L. Buchman, Mace Vaughn, Scott Hoffman Black. *Pollinator Conservation Handbook.* The Xerces Society, 2003.

Simon, Richard. "Deficit Revised Lower for '06, with Bump in '07." *Press Democrat*, August 18, 2006, p. A6.

Stiglitz, Joseph. "The Tyranny of King Cotton." *Atlanta Journal-Constitution*, October 24, 2006.

Tesconi, Tim. "Could Global Warming Dry Up Wine Industry?" *Press Democrat*, July 11, 2006, pp. A1, A11.

Tilman, David. "The Greening of the Green Revolution." *Nature*, 19 November 1998, pp. 211–12.

Trauner, Carol. "Leaves of Grass: The Growing Popularity of Grass-fed Beef and What You Should Know About It." Chefs Collaborative Communiqué, November 2003.

Weiss, Rick. "Gene-Altered Profit-Killer: A Slight Taint of Biotech Rice Puts Farmers' Overseas Sales in Peril." *Washington Post*, September 21, 2006.

Wellman, Nancy S., and Barbara Friedberg. "Causes and Consequences of Adult Obesity: Health, Social and Economic Impacts in the United States." Asia Pacific Journal of Clinical Nutrition 11 (2002): S705–S709, web edition.

Wen, Dale. "China Copes with Globalization: A Mixed Review." International Forum on Globalization, 2005.

Wilkins, Jennifer. "The Oil We Eat: Fossil Fuels Consume Big Portion of Food Costs." TimesUnion.Com, May 7, 2006.

Williams, Ted. "Salmon Stakes." *Audubon*, March 2003, online edition.

Williamson, Elizabeth. "Some Americans Lack Food, But USDA Won't Call Them Hungry." *Washington Post*, November 16, 2006, p. A01, web edition.

Winne, Mark. "Growing a Healthy Food System–Food and Agriculture in New Mexico." New Mexico Food and Agriculture Policy Council, January 2005.

Wise, Timothy. "Identifying the Real Winners from U.S. Agricultural Policies." Working Paper No. 05-07, Global Development and Environment Institute, Tufts University, 2005.

Womach, Jasper. "Previewing a 2007 Farm Bill." CRS Report for Congress, Order Code RL 33037, Congressional Research Service.

Notes

1. Somewhere in America...

1. There are 56 pounds in a bushel; the United States produces 10 billion bushels of corn per year, up from 4 billion in 1970, Michael Pollan, *The Omnivore's Dilemma: A Natural History of Four Meals,* Penguin Press, 2006, p. 62.

2. The United States and Mexico share a 1952 mile border that spans 4 states. In 2006, the U.S. employed approximately 10,000 border patrol agents.

3. An estimated 1.2 million corn farmers and 400,000 sugar producers have been driven from the land since the signing of the North American Free Trade Agreement due to U.S. dumping of corn and high-fructose corn syrup. See Christine Ahn with Melissa Moore and Nick Parker, "Migrant Farmworkers: America's New Plantation Workers," *Backgrounder* Newsletter, Spring 2004, web version, http://www.foodfirst.org/node/45. According to Tom Philpott, "There's a saying in Mexico that for every bushel of corn you dump on us, we'll send you ten workers," cited in *Daily Grist,* "Toward a green agenda on immigration," April 12, 2006.

4. Cellulosic ethanol entails fermentation and distillation from the cellulose of plant fibers rather than the processed sugars, resulting in positive net energy balances. This will be addressed in a later chapter.

5. Mad-cow disease is also known as bovine spongiform encephalopathy or BSE.

6. According to Michael Pollan, atrazine is a powerful herbicide applied to 70 percent of the nation's corn crop. Concentrations as low as 0.1 part per billion are frequently found in waterways and wells, enough to "chemically emasculate a male frog, causing its gonads to produce eggs." (Pollan, "Mass Natural," *New York Times,* June 4, 2006.)

7. Food miles are an account of the total distance a food item's ingredients travel from farm to table.

2. Why the Farm Bill Matters

8. Essayist and farmer Wendell Berry has written: "The global 'free' market is free to the corporations precisely because it dissolves the boundaries of the old national colonialisms, and replaces them with a new colonialism without restraints or boundaries. It is pretty much as if all the rabbits have now been forbidden to have holes, thereby 'freeing' the hounds." (from "The Total Economy," in *Citizenship Papers,* 2003.)

9. According to Congressman Ron Kind, (D-WI), most people in agriculture realize that changes have to be made, and that current farm policy is distorting the real market. He contends that there are a lot of acres being planted to a few commodities. "The only reason they're planting that is for the government paycheck, not because of the marketplace."

3. What Is the Farm Bill?

10. Jasper Womach, Coordinator, "Previewing a 2007 Farm Bill," Congressional Research Service, Order Code RL33037, August 18, 2005, CRS-1.

11. Ferd Hoeffer, Sustainable Agriculture Coalition, "Farm Bill Primer," PowerPoint Presentation, September 2005.

12. The American farm bloc lobby is politically secure thanks to a relative overrepresentation of rural America in the Senate and tens of millions of dollars in political donations each election cycle. The food and nutrition programs–with backing from urban representatives–historically have provided the "critical mass" of political support for the omnibus Farm Bill from outside of the Farm Belt states.

13. "The Charm of the Farm," Editorial, *Washington Post,* October 19, 2005, web edition.

14. Stephen J. Brady, "Highly Erodible Land and Swampbuster Provisions of the 2002 Farm Act," in *Fish and Wildlife Benefits of Farm Bill Conservation Programs,* 2000–2005 Update, USDA Natural Resources Conservation Service and Farm Service Agency, p. 8.

15. According to Charles Benbrook's analysis, the U.S. ranks 23 out of 34 countries, spending $2.28 per person for each 1000 calories consumed. The countries that spent the least for every 1000 calories consumed were Sierra Leone, 39 cents; Mali, 46 cents; Tanzania, 51 cents; and Kenya, 63 cents. The countries that spent the most per 1000 calories consumed were Korea, $4.43; Japan, $3.68; Argentina, $3.47; Australia, $3.28; and the United Kingdom, $2.96. Americans buy lots of convenience, packaging and services with their food dollars, and as a result, pay a lot more for it.

16. Katherine M. Flegal, et al., "Prevalence and Trends in Obesity Among U.S. Adults, 1999–2000," *Journal of the American Medical Association,* 288, no. 14, October 2002.

4. Promises Broken: The Two Lives of Every Farm Bill...

17. According to one long-time Farm Bill observer, about one-third of the members of the agriculture committees are new representatives that have been assigned to the task and are eager to be released as soon as possible.

18. Alan Guebert reports, for example, that as chairman of the Appropriations subcommittee on agriculture, Henry Bonilla (R-TX), "raked in $250,414 of his $1.05 million in 2001 and 2002 PAC money from agribusiness."

19. Known also as the Harkin plan, after Iowa Democratic Senator Tom Harkin, chairman of the Senate Agriculture Committee, who championed the program in an effort to reform the problems of subsidies and monoculture factory farming.

20. Defenders of Wildlife's Comments for the Development of USDA Recommendations for the 2007 Farm Bill, June 17, 2005

5. Where It All Started

21. Quoted in "Ownership Matters: Three Steps to Ensure a Biofuels Industry That Truly Benefits Rural America," David Morris, based on a speech to the Minnesota Ag expo in Morton, Minnesota, January 25, 2006.

22. Elizabeth Corcoran, "The Answer on the Wind," *San Francisco Chronicle*, Sunday, January 8, 2006, p. M1.

23. Quoted in "Self-Help in Hard Times," in Howard Zinn, *A People's History of the United States*, 389.

24. The longest period of relative parity between farm prices and manufacturing prices was in the years leading up to World War I (1911–1914), and this was used to determine the price parity index. See John C. Culver and John Hyde, *American Dreamer: The Life and Times of Henry A. Wallace*, W.W. Norton, New York, 2000, p. 56 for an extended analysis.

25. Culver and John Hyde, *Ibid.*, p. 99.

26. Michael Pollan, *The Omnivore's Dilemma*, p. 48.

27. Bruce Bartlett, "How Excessive Government Killed Ancient Rome," *Cato Journal*, pp. 289–290.

28. An excellent account of this period can be found in *American Dreamer*.

29. Pollan, Ibid., p. 49.

30. It was estimated that Farm Bill spending during these troubled times had a multiplier effect of seven. That is, for every dollar of government funds spent on farm and food policies, it generated seven dollars in the overall economy. See *American Dreamer* for more details.

31. Richard Manning, *Against the Grain: How Agriculture Hijacked Civilization*, p. 171.

32. Bernard De Vito, quoted in *The King of California*, p. 186.

33. Written in 1924, referenced in *The King of California*, p. 375.

6. Family Farms to Mega-Farms

34. The Soviets, with the cooperation of large grain companies, quietly purchased large amounts of grain at pre-inflationary prices in the early 1970s. Prices soared after the Russian Grain Deal was announced, but most farmers had already sold their grain at low prices.

35. One of the most notorious incidents of this era involved the shooting of banker Rudy Blythe and chief loan officer Deems Thulin by James and Steven Jenkins in Ruthton, Minnesota, in 1983. The bank had foreclosed on the father and son's dairy operation and James and Steven Jenkins took their rage out on unsuspecting bank officials. See "Twenty Years After the Ruthton Banker Killings: Desperation Still Simmers," Paul Levy, *Star Tribune*, Monday, October 20, 2003.

36. Michael Pollan, *The Omnivore's Dilemma*, pp. 52–53.

37. Ibid.

38. "Trade Reforms and Food Security," Food and Agriculture Organization, Rome, Italy, 2003.

39. Mary Hendrickson and William Heffernan, "Concentration of Agricultural Markets," February 2005, Department of Rural Sociology, University of Missouri, web edition.

40. Mark Drabenstott, "Do Farm Payments Promote Rural Economic Growth?" *Main Street Economist*, March 2005, Federal Reserve Bank of Kansas City.

7. The Farm Bill's Hunger Connection

41. Dennis Roth, "Food Stamps: 1932–1977: From Provisional and Pilot Programs to Permanent Policy," Economic Research Service.

42. Ibid., p. 8.

8. The Conservation Era Begins—Again

43. Ronald Reynolds, "The Conservation Reserve Program and Duck Production in the U.S. Prairie Pothole Region," in *Fish and Wildlife Benefits of Farm Bill Conservation Programs, 2000-2005 Update*, USDA NRCS and Farm Service Agency, p. 35.

44. The 2002 Farm Bill provided funding for an additional 1,000,000 acres of wetland set asides and restoration, mostly in the Southeastern bottomland forests that should never have been farmed in the first place.

45. "The lower 48 states had an estimated 220 million acres of wetlands and streams in precolonial times, but 115 million acres of them had been destroyed by 1997." In "U.S. reports increase in wetland acreage: Bush Administration figures are disputed as being misleading," John Heilprin, Associated Press, Friday March 31, 2006, *San Francisco Chronicle*, p. A2.

46. Over 80 percent of species use aquatic habitats at some point in their life cycle. Creek corridors are probably the single most important wildlife linkages, as they connect all other habitats and lie at the heart of an ecosystem.

47. Congressman Ron Kind's "Healthy Farms, Foods, and Fuels Act of 2006" calls for a doubling of water protection incentives to $2 billion per year and a restoration of 3 million acres of wetlands.

48. 60 percent of EQIP funds were committed to livestock operations.

49. Suzie Greenhalgh, Mindy Selman, and Jenny Guiling of the World Resources Institute quote the following sources for Conservation program spending: (1) EQIP funding allocation: www.nrcs.usda.gov/programs/eqip/; Personal communication: Edward Brzostek (USDA Natural Resources Conservation Service, June 2006; (2) Conservation Reserve Program funding allocation 29th signup: www.fsa.usda.gov/dafp/cepd/29th/TheConservationReserveProgram29thSignup.pdf. (3) Grassland Reserve Program funding allocation: www.nrcs.usda.gov/programs/grp/; (4) Wetlands Reserve Program funding allocation: www.nrcs.usda.gov/programs/wrp/; (5) Wildlife Habitat Incentives program funding allocation: www.nrcs.usda.gov/programs/whip/; personal communication: Albert Cerna (USDA-NRCS), June 2006.

50. According the the World Resources Institute, "in 2005, USDA NRCS changed operating systems for EQIP applications and NRCS suspects that many states were not able to submit all their applications because of the workload to migrate data from the old system to the new system. Therefore, the percent of applications funded in 2005 may seem artificially high. (Personal communication: Edward Brzostek, USDA-NRCS, June 5, 2006)."

51. FAO Newsroom, "Livestock a Major Threat to Environment: Remedies Urgently Needed," November 29, 2006, web edition.

9. Freedom to Farm and the Legacy of Record Payoffs

52. Scott Marlow, "The Non-Wonk Guide to Understanding Federal Commodity Payments," The Rural Advancement Foundation International–USA, 2005 edition, page 3.

53. Ibid.

54. Dan Morgan and Gilbert Gaul, "No Farm? No Crop? No Problem–Subsidies Still Grow," *Washington Post*, July 2, 2006, web edition.

55. According to the Sustainable Agriculture Coalition, 400,000 CRP contracts become eligible for renewal between 2007–2010. Many are not being renewed and instead are being plowed up for commodity production.

10. Who Gets the Money?

56. Peanuts and tobacco were long-time stalwarts but have been phased out through payout programs.

57. For an interesting study that analyzes the ties between subsidies and county economics, see "What Would Happen if Federal Farm Subsidies Were Eliminated? Evidence for Colusa and Tulare Counties," Sandra Gonzlez, Rachael Goodhue, Peter Berck, and Richard Howitt, Giannini Foundation of Agricultural Economics, web version at www.agecon.ucdavis.edu/uploads/update_articles/v8n5_4.pdf.

58. For a detailed analysis, read Mark Arax and Rick Wartzman, *The King of California: J. G. Boswell and the Making of a Secret American Empire*, Public Affairs, 2003.

59 Environmental Working Group, "After Hong Kong, Redraw America's Subsidy Map," December 13, 2005, http://www.ewg.org:16080/farm/redraw/.

60. Ibid.

61. Dan Chapman, Ken Foskett, and Megan Clarke, "How savvy growers can double, or triple, subsidy dollars," The *Atlanta Journal-Constitution*, Oct. 2, 2006

62. Dan Chapman, Ken Foskett, and Megan Clarke, "How your tax dollars prop up big growers and squeeze the little guy," *The Atlanta Journal-Constitution*, Oct. 1, 2006.

63. Forrest Laws, "GAO Report Sheds Little Light on Payment Limit Rules," Southwest Farm Press, July 1, 2004.

64. Morgan and Gaul, "No Farm? No Crop? No Problem–Subsidies Still Grow," *Washington Post*, July 2, 2006. These reporters specifically mention Mary Anna Hudson, from River Oaks near Houston who received $191,000, and Houston surgeon Jimmy Frank Howell, who received $490,709 since the 1996 Freedom to Farm bill.

11. The Multiple Benefits of Skilled Farmers and Healthy Rural Lands

65. The Millennium Ecosystem Assessment provides a cogent analysis, www.maweb.org/en/index.aspx.

12. World Trade Organization Rulings and the Era of Green Payments

66. Peter Rosset, "Giving Away the Farm: The 2002 Farm Bill," Institute for Food and Development Policy, Oakland, California, Summer 2002.

67. Joseph Stiglitz, "King cotton's tyranny," *Atlanta Journal Constitution*, October 8, 2006.

68. U.S. cotton farmers receive an average of $230 per acre in subsidies.

69. Dan Chapman, Ken Foskett, and Megan Clarke, "How Your Tax Dollars Prop Up the Big Guys and Squeeze the Little Guy," *Atlanta Journal-Constitution*, October 1, 2006.

70. Agriculture Statistics, 1992. Statistics of Cotton, Tobacco, Sugar Crops and Honey. Table 81, p. 61.

71. Agriculture Statistics, 2002. Statistics of Cotton, Tobacco, Sugar Crops and Honey. Table 2-1, page II-1.

72. Environmental Working Group Subsidy Data Base.

13. New Zealand: Still Subsidy-free After All These Years

73. John Pickford, "New Zealand's Hardy Farm Spirit," BBC, New Zealand, October 16, 2004, web edition.

74. Ibid.

75. Laura Sayre, "Farming Without Subsidies? Some Lessons from New Zealand," *The New Farm*, March 2003.

76. Ibid.

77. Ibid.

15. Hunger, Health, and Nutrition: Changing the Policy Palette

78. "U.S. Farm Policy Contributes to Obesity: New Report Finds Policies Drive Production of Unhealthy Foods," Institute for Agriculture and Trade Policy, 2006, p. 5.

79. Food Insecurity means that a household had limited or uncertain availability of food, or limited or uncertain ability to acquire acceptable foods in socially acceptable ways (i.e., without resorting to emergency food supplies, scavenging, stealing, or other unusual coping strategies).

80. "Out of Balance: Marketing of Soda, Candy, Snacks, and Fast Foods Drowns Out Healthful Messages," Consumers Union and California Pan-Ethnic Health Network, September 2005.

81. Obesity can be determined by the following formula: weight in kilos (2.2 pounds per kilo)/height in meters (39.9 inches per meter). Obesity occurs if the ratio is greater than 25.

82. The United Kingdom Department of Health's Estimated Average Requirements (EAR) are a daily calorie intake of 1940 calories per day for women and 2,550 for men. These figures apply to adults with low activity levels.

83. Gary Paul Nabhan, *Why Some Like It Hot: Food, Genes, and Cultural Diversity*, Island Press, 2004, pp. 175–177.

84. Mark Muller and Heather Schoonover, "Food Without Thought: How U.S. Farm Policy Contributes to Obesity, Institute for Agriculture and Trade Policy, 2006, p. 7

85. Ibid., p. 6.

86. Marion Nestle, "One Thing to Do About Food," *The Nation*, September 11, 2006, p. 14.

87. Muller and Schoonover, p. 8.

88. "Making the Case for Local Food Systems," A Farm and Food Policy Project Learning Paper, p. 2.

89. Janet Raloff, "Money Matters in Obesity," *Science On-Line*, July 16, 2005, vol. 168, no. 3.

90. Farm and Food Policy Project, "Making the Case for Local Food Systems," Learning Paper, October 2006.

91. Farm and Food Policy Project, "Making Healthy Food More Accessible for Low-Income People," Learning Paper, October 2006.

92. Ibid.

93. Nancy Wellman and Barbara Friedberg, "Causes and consequences of adult obesity: health, social and economic impacts in the United States," *Asia Pacific Journal of Clinical Nutrition*, (2002) web edition.

94. Scott Carlson, Globe-trotting for Groceries, *Utne Reader*, March–April 2006, p. 83.

95. Alan Guebert, "Special Interests Gut COOL Funding," *The New Farm*, Monday, June 23, 2003.

96. "Country of Origin Labeling Laws for Food," Food Management Institute.

97. Scot Kilman, "U.S. Food Giants Fight Country of Origin Labeling: Grocers, Meatpackers Fight Law to Label Origin of Food Products," *Wall Street Journal*, June 26, 2003.

98. Carlson, Ibid.

99. Ibid.

100. Ibid.

101. Ibid.

102. Carlson, Ibid.

103. Guebert, Ibid.

104. Carlson, Ibid.

16. Energy: Farming in an Era of Fossil Fuel Scarcity

105. Other obvious potential drivers include scarce fresh water supplies, health and nutrition, global warming, and rising national debt.

106. Peter Huber and Mark Mills, *The Bottomless Well: The Twilight of Fuel, the Virtue of Waste, and Why We Will Never Run Out of Energy*, Basic Books, 2005, p. 7.

107. Pollan, *The Omnivore's Dilemma.*

108. Richard Manning, "The Oil We Eat," *Harper's* Magazine, February 2005.

109. Pollan, *Ibid.*, p. 84.

110. Author Michael Pollan cites a Cornell scientist's estimate that growing, processing, and shipping one calorie's worth of arugula from California to the East Coast costs fifty-seven calories of fossil fuel in Steven Shapin, "Paradise sold: What are you buying when you buy organic," *New Yorker* magazine, May 15, 2006.

111. Johannes Kotschi and Karl Müller-Sämann, "The Role of Organic Agriculture in Mitigating Climate Change: A Scoping Study," International Federation of Organic Agriculture Movements, May 2004.

112. According to a USDA Renewable Fuels and Products Information Sheet, The LHV (lower heating value) of ethanol is 11,500 BTU/lb., or 75,700 BTU/gallon. Gasoline is 19,000 BTU/lb., or 115,500 BTU/gallon.

113. *Consumer Reports*, "The Ethanol Myth: Consumer Reports' E85 Tests Show That You'll Get Cleaner Emissions but Poorer Fuel Economy...If You Can Find It," October 2006.

114. Ibid.

115. David Morris, "The Carbohydrate Economy, Biofuels, and the Net Energy Debate," The Institute for Local Self-Reliance, 2005.

116. John Deutch, "Biomass Movement," *Wall Street Journal*, May 10, 2006.

117. It is important to note that petroleum is also heavily subsidized, with huge tax breaks, military campaigns, costs related to climate change and other environmental, social, and health impacts.

118. John Deutch, "Biomass Movement," *Wall Street Journal*, May 10, 2006, p. A18.

119. Tom Philpott, "Archer Daniels Midland: The Exxon of Corn," *The Daily Grist*, February 2, 2006.

120. USDA biofuels expert, personal communication.

121. Joe Millman, "Labor Movement: As U.S. Debates Guest Workers, They Are Here Now in Construction 'Subidos,'" *Wall Street Journal*, September 18, 2006, A1.

122. A process known as CR^3, steams municipal waste (a continual source heavy in paper and cardboard) into a variety of useful products.

123. Deutsch, Ibid.

124. *Consumer Reports*, Ibid.

125. Liz Marshall and Suzie Greenhalgh, "Beyond the Renewable Fuels Standard: The Environmental and Economic Impacts of Grain Ethanol Production in the United States," World Resources Institute, September 2006, p. 5.

126. See David Morris, "The New Ethanol Fuel Demands a New Public Policy," Institute for Local Self-Reliance. Morris argues that there is such a precedent in the royalty paid to oil companies for off-shore drilling that fluctuates with the market price for crude oil.

127. Environmental Law and Policy Center, "An American Success Story: The Farm Bill's Clean Energy Programs."

128. Kotschi and Müller-Sämann, Ibid.

129. Andrew Pollack, "Redesigning Crops to Harvest Fuel," *New York Times*, Friday, September 8, 2006, p. C1-C-4.

130. Ibid.

131. Ibid.

132. Ibid.

17. Healthy Lands, Healthy People

133. Jerry Goodbody, "Green Acres," *Audubon* Magazine, November 2005.

134. Defenders of Wildlife's Comments for the Development of USDA Recommendations for the 2007 Farm Bill (70 Federal Register 35221 June 17, 2005).

135. Jay Chamberlain, California Resources Agency, Presentation at California Food and Farming Coalition public forum, January 18, 2006.

136. Sustainable Agriculture Coalition, "No Time for Delay: A Sustainable Agriculture Agenda for the 2007 Farm Bill," p. 10.

137. 60 percent of EQIP moneys must be spent on livestock operations. According to the NRCS, large feedlot operations receive about 25 percent of all EQIP dollars. Feedlots can be compensated up to $450,000 for the construction of manure lagoons and animal waste processing facilities. Equating basic environmental compliance with conservation is quite a stretch.

138. Sustainable Agriculture Coalition, "No Time for Delay," p. 49.

139. Arthur Allen addresses this in "The Conservation Reserve Enhancement Program," in *Fish and Wildlife Benefits of Farm Bill Conservation Programs, 2000–2005 Update*, NRCS, p. 123.

140. Stephen J. Brady, "Highly Erodible Land and Swampbuster Provisions of the 2002 Farm Act," *Fish and Wildlife Benefits of Farm Bill Conservation Programs 2000–2005 Update*, p. 8.

141. Ronald E. Reynolds, "The Conservation Reserve Program and Duck Production in the U.S. Prairie Pothole Region," *Fish and Wildlife Benefits of Farm Bill Conservation Programs 2000–2005 Update*, pp. 34–35

142. Brady, Ibid., p. 12.

143. "No Time for Delay," Sustainable Agriculture Coalition, October 2006, p. 50.

144. Ibid., p. 50.

145. Private interview.

146. Douglas Johnson, "Grassland Bird Use of Conservation Reserve Program Fields in the Great Plains," in *Fish and Wildlife Benefits of Farm Bill Conservation Programs, 2000–2005 Update*, NRCS, p. 26.

147. Matthew Shepherd, Stephen Buchman, Mace Vaughn, and Scott Hoffman Black, *Pollinator Conservation Handbook*, Xerces Society, 2003, p. 6.

148. Ted Williams, "Salmon Stakes," *Audubon* Magazine, March 2003.

18. Turning the Tables

149. See the *Madison Area Community Supported Agriculture Coalition* (MACSAC) and Physicians Plus Eat Healthy Rebate Program for more information.

150. Kirsten Ferguson, "In Profile: Preston Maring, M.D.," *American Farmland*, American Farmland Trust, Summer 2007, web edition.

151. Cohen et al., 1997.

152. Anna Lappé, "Doing Lunch: Ann Cooper Serves Up a New Vision of School Food," *The Nation*, September 11, 2006, p. 35.

153. Ibid.

154. Brian Halweil, "Can Organic Farming Feed Us All?" *WorldWatch*, May/June 2006.

155. Ibid.

156. Kotschi and Müller-Sämann, 2004.

157. James Randerson, "Organic farming boosts biodiversity," *New Scientist*, web edition.

158. Becky Weed, "A Grassland Manifesto," in *Farming and the Fate of Wild Nature: Essays in Conservation-based Agriculture*, Watershed Media, 2006, p. 88.

Photography Credits

First Color Section

Page i: Girl carrying U.S. food aid, Cameroon, © Mark Edwards/Peter Arnold, Inc

Page ii: Illegal entry at U.S./Mexico border, Tijuana, Mexico, © Lenny Ignelzi/AP Images

Page iii: Farm, NRCS

Page iv: Fish kill at Klamath River, California, © Ron Winn/AP Images

Page v: Three Mile Canyon Farm, Oregon, © Dan Imhoff

Page vi: Supermarket shoppers © Scott Vlaughn

Page vii: Homeless man, San Francisco, © Mark Downey/ Painet, Inc.

Page viii: School lunch tray, © Jared Lawson, courtesy of the Community Food Security Coalition

Main Text

Page 31: Sierra Orchards, Winters, California, © Dan Imhoff

Page 32/33: Migrant mother, Nipomo, California, 1936 © Dorothea Lange

Page 37: Coon Creek Watershed, USDA; inset photo, © Dan Imhoff

Page 41: Combines, NRCS

Page 44: Kennedy, West Virginia, © Cecil Stoughton/JFK Presidential Library

Page 45: "We Need Food", © Martha Tabor

Page 51: Manure lagoon, NRCS

Page 52: Manure lagoon, NRCS

Page 66: Girl carrying U.S. food aid, Cameroon, © Mark Edwards/Peter Arnold, Inc

Page 69: Streamside restoration, Antietam watershed, Maryland, © Dan Imhoff

Page 71: Mailbox, © Roberto Carra

Page 76: WTO protest, Mexico, © Laura Rauch/AP Images

Page 77: Cotton, NRCS

Page 78: Cotton Harvest, Eritrea, © Jorgen Schytte/Peter Arnold, Inc.

Page 79: Cotton, NRCS

Page 81: Sheep, NZ Ministry of Agriculture and Forestry

Page 94: Supermarket shopper, © Scott Vaughn

Page 101: Shopping for meat, © Richard Levine/ Alamy Images

Page 102: Slaughterhouse, USDA

Page 112: Windmill, © Doug Tompkins

Page 126: Farm, Iowa, 1999 © Lynn Betts

Page 127: Animus Valley, NM © Dan Imhoff

Page 133: Urban sprawl, Great Lakes, © Lynn Betts/NRCS

Page 136: NRCS

Page 137: Greg & Kathy Koether, © Brian DeVore

Page 138, 145: School lunch program, Thousand Oaks Elementary School, © Craig Lee

Final Color Section

Page i: Hayfield, © Gary Randorff
Bison; Prairie Potholes © Roberto Carra
Sandhill cranes in rice field © Robert Payne

Page ii-iii: Canal restoration © John Anderson
Bobcat © Sue Morse
Cornfield Willamette Valley Oregon; Elk at water's edge © Roberto Carra
Llama and sheep at 13-Mile Farm; Benzinger Vineyards Insectory © Dan Imhoff
Hayride © Dan Guenther

Page iv-v: Free-range cows; Solar panel array © Dan Imhoff
Sunflowers; Cowboys; Apple picking © Roberto Carra
Bumble bee © David L. Green
Potatoes © Doug Gosling

Page vi-vii: Greenhouse © Barbara Damrosch
Harlem Garden © Peter Forbes
Farmer at farmer's market; workers at community garden © Jared Lawson
Girl at farmer's market © Scott Vlaughn
Oranges © Roberto Carra

Page viii: Student with salad © Jared Lawson
Student restoration project © Audubon California
Greens © Scott Vlaughn
Los Angeles community garden © Jared Lawson

Afterword

Somewhere in America's Future...

It is not that difficult to imagine a time in America's future when the sun rises over vastly different agricultural landscapes. Out of vision, and out of necessity, citizens will begin to see and value food and farm policy as part of a much larger orbit of social, economic, and environmental concerns. Government support for food and agriculture may remain significantly high, (and may even exceed current levels), but spending programs will pass rigorous tests for costs and benefits. Of course, imagining is the easy part. Getting there may require moving mountains.

In such a future, citizens, consumers, and food producers will understand they are bound by similar fates. The nutrition of the body reflects the health of the land. American farmers and ranchers will produce an abundance of some of the finest crops and livestock in the world, but they will be more fairly rewarded for their efforts. And the farms, ranches, and forests that cover nearly two thirds of the contiguous United States will supply far more than food. With incentives for proper management, they will also provide clean air and water, renewable energy, wildlife habitat, diverse forests, and open space.

Healthy, locally produced foods will form the basis of a modern national health care prevention strategy. With ever-escalating fuel prices and other concerns, the need for relative autonomy in food production will become an organizing principle in nearly every region of the country. In public schools, children will have direct contact with numerous farms that provide their cafeterias with grass-fed milk and meat, eggs, organically raised fruits, vegetables, and grains. A similar transformation will sweep hospitals and universities, corporate campuses, and government agencies. Obesity rates will eventually begin to decline, while the costs of Medicare, Medicaid, and public health fall. Worker productivity will rebound.

Traveling through the countryside, one will see that a new vision has taken hold. Monoculture fields that once blanketed entire counties and regions instead include large areas of perennial grasses, restored prairies, and cover crops. Wooded field margins and vibrant creek banks transect row crop and orchard operations. Large set aside areas of protected wildlands and natural habitats serve as buffer zones against extreme storm events. Organic agriculture, a preferred farming method in many regions, requires more people than strategies used in agriculture at the turn of the twenty-first century, but also reduces energy inputs and harmful air emissions and raises the nutritional content of foods. Greenhouses extend growing seasons

for fresh produce categories that can withstand cooler temperatures, such as salad greens and root crops.

No single reform will more dramatically transform the landscape than the large-scale conversion from confinement animal feedlot operations to diversified farms that include grass-pastured livestock. Under a national grassland recovery campaign, grass farmers will become emblematic of a new family farm movement. On-farm generated incomes will also begin to rise. Soil erosion will significantly stabilize while agricultural runoff and farm-related water pollution decreases. Grassland bird species will become common as nesting habitats return. Herds of bison even return to vast areas of the Great Plains that for decades were fragmented by artificially green crop circles. Formerly threatened species such as the sage grouse and prairie chicken revive because of collaborative stewardship efforts.

"Food deserts" lacking in fresh locally grown fruits and vegetables will still exist in intensive agricultural production zones, though healthy food oases will have also sprung up across the country in rural and urban areas. Much of this will be made possible through a new generation of supply networks. Regional economies will emerge as powerful supporters of family farmers, with a variety of innovative methods for storing, transporting, processing, and distributing locally grown foods. The Food Stamp Program will remain a front line of

INSECTORY

defense against hunger and food insecurity. Yet it too will evolve through a heightened awareness that delivering not just calories, but nutritious foods to those in need best serves the long-term interests of citizens and the country alike. Thanks to nearly heroic community organizing and the difficult work of forging of public and private partnerships, access to healthy foods will expand into neighborhoods most at risk for nutritionally preventable diseases such as hypertension, obesity, and diabetes.

Such a transformation will only have been made possible by a transformation inside the Washington D.C. beltway. Food and farm policy, still subject to politicking and opportunism, becomes less of a Matrix-like parallel universe in which the corporations control and dominate the subsidy system in the name of the small farmer. The Food and Farm Bills that lawmakers debate and reauthorize every five to seven years will be more transparent and accessible to the average citizen. Research and development efforts will focus on finding solutions to urgent goals facing the country: decreasing energy inputs, reducing global warming emissions, protecting declining wildlife species, and encouraging the next generation to take up the challenge. The United States will find itself in a leadership role in the development of innovative small- and medium-scale methods and technologies that produce food, fiber, feed, and energy in response to ongoing environmental

challenges and ever-changing conditions and awareness.

Looming problems will persist, including food insecurity for some, and climate changes that make agriculture and conservation even more unpredictable and vulnerable. The tug of war between global economics, the inclination to farm industrially, a continually growing population, and the fragility of wild nature will continue. We will still eat too many French fries. But a chapter will have been closed and a new one begun. The critical moment will occur when a coalition of previously isolated voices joins ranks to challenge the status quo by insisting on a healthier, more hopeful and secure future for themselves, their children, and grandchildren. Farm and food policy will become an integrated economic engine that not only encourages environmentally viable crop production but truly supports health and nutrition, renewable energy, entrepreneurial development, stewardship, fair trade, living wages, and regional food security.

TATSOI